Poultry Health and Management

Chickens, Turkeys, Ducks, Geese, Quail

Other books of interest

Farm Machinery
12th Edition
Claude Culpin
0 632 03159 X

Primrose McConnell's The Agricultural Notebook
18th Edition
Edited by R.J. Halley and R.J. Soffe
0 632 03474 2

Fream's Principles of Food and Agriculture
17th Edition
C.R.W. Spedding
0 632 02978 1

Aquaculture – Principles and Practices
T.V.R. Pillay
0 85238 202 2

Dalton's Introduction to Practical Animal Breeding
3rd Edition
M.B. Willis
0 632 03126 3

Intensive Fish Farming
Jonathan Shepherd and Niall Bromage
0 632 03467 X

Crustacean Farming
Daniel O'C. Lee and John F. Wickins
0 632 02974 9

Poultry Health and Management

Chickens, Turkeys, Ducks, Geese, Quail

DAVID SAINSBURY
BSc, MA, PhD, MRCVS, FRSH, FIBIOL
Fellow of Wolfson College and
Lecturer in Animal Health,
Department of Clinical Veterinary Medicine,
University of Cambridge

THIRD EDITION

OXFORD
BLACKWELL SCIENTIFIC PUBLICATIONS
LONDON EDINBURGH BOSTON
MELBOURNE PARIS BERLIN VIENNA

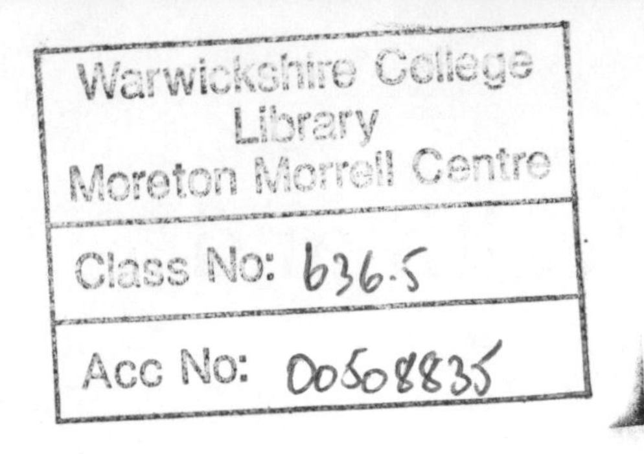

Blackwell Scientific Publications
Editorial offices:
Osney Mead, Oxford OX2 0EL
25 John Street, London WC1N 2BL
23 Ainslie Place, Edinburgh EH3 6AJ
238 Main Street, Cambridge
Massachusetts 02142, USA
54 University Street, Carlton
Victoria 3053, Australia

Other Editorial Offices:
Librairie Arnette SA
2, rue Casimir-Delavigne
75006 Paris
France

Blackwell Wissenschafts-Verlag GmbH
Düsseldorfer Str. 38
D-10707 Berlin
Germany

Blackwell MZV
Feldgasse 13
A-1238 Wien
Austria

First published by
Granada Publishing 1980
Reprinted 1982
Second edition 1984
Reprinted by BSP Professional Books 1987
Third edition by
Blackwell Scientific Publications 1992
Reprinted 1993

Set by DP Photosetting, Aylesbury, Bucks
Printed and bound in Great Britain by
Hartnolls Ltd, Bodmin, Cornwall

DISTRIBUTORS

Marston Book Services Ltd
PO Box 87
Oxford OX2 0DT
(*Orders:* Tel: 0865 791155
Fax: 0865 791927
Telex: 837515)

USA
UNIPUB
4611-F Assembly Drive
Lanham, MD 20706
(*Orders:* Tel: (301) 459 7666
(800) 274 4888
Fax: (301) 459 0056)

Canada
Oxford University Press
70 Wynford Drive
Don Mills
Ontario M3C 1J9
(*Orders:* Tel: 416 441-2941)

Australia
Blackwell Scientific Publications Pty Ltd
54 University Street
Carlton, Victoria 3053
(*Orders:* Tel: 03 347-5552)

British Library
Cataloguing in Publication Data
A catalogue record for this book is available from the British Library

ISBN 0–632–03325–8

Library of Congress
Cataloging in Publication Data

Sainsbury, David.
Poultry health and management: chickens, turkeys, ducks, geese, quail / David Sainsbury. — 3rd ed.
Includes bibliographical references and index.
ISBN 0–632–03325–8
1. Poultry. 2. Poultry—Health.
I. Title.
SF487.S15 1992 92-23280
636.5—dc20 CIP

Contents

Preface to First Edition

Preface to First Edition

It appears as a surprising fact that, in spite of the enormous and exciting technical strides made in poultry husbandry and health in recent years there are remarkably few texts that bring the practical aspects of this together so that all those concerned either with the keeping of poultry or eager to learn about it can find much that they require under one cover.

My own work brings me into close contact with poultry units and their problems, both in the United Kingdom and abroad, and I also have commitments in teaching both agricultural and veterinary students in the subjects of poultry management and health. This has convinced me of the need for a book that deals concisely with the application of our overall scientific knowledge to the practical management of poultry.

It may be helpful if I explain my approach to dealing with poultry management in this work. In the first place I have explained briefly the overall structure of the industry and how it has tended to concentrate on a limited number of systems for each type of bird. It is also shown how great has been the increase in efficiency in recent years with the poultry industry very much in the van of all agricultural developments. Following this introductory section, which includes an explanation of the breeding policies of the hybrid poultry companies, there are explanations of the nutritional needs of all classes of poultry.

The importance of poultry feeding cannot be over-emphasised and I hope this has been dealt with sufficiently as a knowledge of this is fundamental to good management. So also is a complete understanding of the environmental needs of poultry and the ways in which they may be achieved in a practical manner. Thus I deal in these sections with the air temperature, lighting, and ventilation needs of poultry, together with the construction and thermal insulation of the poultry house. These are all items of management which are very much under the control of the poultryman himself and these sections have been written to give him the underlying knowledge to deal with these aspects.

In the chapter on 'Disease and intensification', my philosophy is explained in that every effort must be concentrated into the prevention of disease and with a good understanding of the factors that cause disease, healthy poultry are a practical reality!

The subject of poultry health forms a major part of this book; so far as possible I have attempted to avoid the 'dictionary of disease' or 'pathological' approach, which is much more for the specialist.

I have explained the principal systems of poultry management and even those currently less popular since there is plenty of evidence that most systems of management have a part to play somewhere in the world. Poultry husbandry has become a highly international matter; for example, broiler chickens are reared in almost precisely the same way in the U.S.A. as the U.K., or in Africa, Asia or the Middle East. This is meant to be a book for international use which can enable the poultry keeper to obtain pleasure and profit from rearing productive and healthy stock. Vast waste takes place throughout the world due to the inefficient use of our resources and it is our duty to make certain that when poultry are kept, their great potential productivity is thoroughly ensured by their correct management.

Fairly brief sections have been given on the husbandry of turkeys and ducks, emphasising in particular those aspects which are different from the domestic fowl. The consideration given in the earlier chapters is not, however, exclusive to the chicken but bears in mind the need of other poultry.

I am not ashamed to admit that I find my involvement with all aspects of poultry keeping over a period of forty years – since in fact I kept laying birds semi-intensively and fattening chicken in individual pens 'Sussex style' from the age of ten – has been an enormously pleasurable one. It does not matter how big or small a unit is, the birds still respond acutely to their management, and if I can instil some of this enjoyment and attitude into the reader, a great deal of the aim of this book will have been achieved. Bibliographies of further reading are given where a deeper understanding can be obtained and I hope the books quoted will be sought out.

David Sainsbury

Preface to Third Edition

There has been no slowdown in the pace of development in the poultry industry and the necessity for a further edition has enabled some major amendments to be made to ensure that the reader is up-to-date with all the changes and trends. The book now includes sections on quail and geese, which are of increasing interest and there is more information on the so-called 'alternative' systems of husbandry which aim to help poultry on more extensive or free-range systems. There have also been major additions to the sections on health and disease, since not only have there been considerable changes in the techniques of disease prevention, but the eruption of the salmonella 'scare' suggested that the public was in danger of ill-health from the consumption of poultry products. It has been vital to deal comprehensively with this issue.

No chapter in the book has gone unaltered, which is clear evidence that the poultry industry is alive and well and continues its growth internationally.

David Sainsbury

Acknowledgements

The greatest pleasure in the production of this book is to express thanks to all those who have given help. Mr James Hunnable and Mr John Baron of the Cobb Breeding Company provided important information and illustrations. Arnold Elson, who is the expert consultant on poultry housing, equipment and systems in the Agricultural Development and Advisory Service (ADAS) has granted me permission to reproduce information and diagrams on alternative systems. His work is especially applauded together with that of his colleagues in ADAS to whom the whole industry is grateful.

My thanks are likewise extended to Mr G.E.S. Robbins for allowing the use of diagrams and material from his books on quail. He is an undoubted expert on the subject and his books are listed in the section on Further Reading. Tables 4.1, 4.2 and 4.3 are reproduced by permission of the Ministry of Agriculture, Fisheries and Food and Table 11.2 by permission of Merck and Company, Rahway, New Jersey. I also thank Mr Peter Fuller Lewis of Pan Britannica Industries for permission to use the illustration of Turbair applications, and Mrs Wyeld for making available tables of rations for ducks. The poultry industry is also served with great distinction by Mr John Farrant, Editor of *Poultry World*, and his colleagues, and Mr Terry Evans and Mr David Martin, Editor and Technical Editor respectively of *Poultry International.*

My further thanks go to Mr West Rose and Mr Graham Hewett of Upjohn Limited, Mr Ralph Auchincloss of Antec International, Mr Robert Bridgeman and Mr Simon Everard of Grampian Pharmaceuticals, Dr Franck Faucomprez of Rhône Mérieux and Mr David Danson of Pitman-Moore. It is impossible to exaggerate the pioneering work of these and other companies who provide the poultry industry world-wide with well researched and manufactured biological and medicinal products.

Finally my grateful thanks to Mr Richard Miles and Mrs Janet Prescott and their colleagues of the publishers Blackwell Scientific Publications, and to my secretary, Miss Peggy McKean, who have all given invaluable and constant support at every stage of this book.

1 Poultry – their health and management – introduction

The background

In very recent times, over a span of no more than thirty years, there have been greater changes in methods of keeping poultry than probably in any other sector of the world's livestock agricultural production. It has now become the most intensive of all branches of livestock farming. In many countries poultry are now ahead of all other livestock in economic importance. Productivity has increased remarkably. Table 1.1 shows how remarkably productivity has improved and systems of management have changed even in as short a time as the last twenty years. In 1960 all systems for the production of eggs recorded an average number of eggs per annum of 185 per bird. Now it is 253. Also, whereas in 1960 there were over 30 per cent of laying poultry still outside on 'free range' and only 19 per cent in cages, now the figure is 13 per cent on 'free range' and 85 per cent in cages. Table poultry are now produced almost entirely intensively and the recent very large increases in fodder costs for poultry will tend to make fully extensive systems almost always uneconomic unless there is a specific demand for products produced in this way and for which a much higher price is paid. Another remarkable fact pertains to table birds: whereas in 1952 it took about 13 weeks to produce a 2.0 kg bird consuming 6.0 kg of food, in 1992 a 2.0 kg bird is grown in approximately 6 weeks at a consumption of 3.8 kg of food.

Intensification

Poultry have undoubtedly led the field in the advanced technical processes required for intensification and indeed the lessons learned by the poultry keeper have often been copied by the rearers of other livestock. It is true to say that any up-to-date guide to poultry health and management can be consulted appropriately by those who keep any farm livestock, as much of the technology, if not the husbandry, is common to all – but with poultry usually the progress leader.

Table 1.1 Yield per laying fowl (eggs per annum – September to August)

System of management	1960/61	1971/72	1972/73	1976/77	1989/90
Free range	166.5	190.9	181.2	192.1	220
Battery	206.2	235.7	238.5	245.4	290
Deep litter and others	187.9	209.2	214.0	224.5	250
All systems	185.1	230.5	234.1	243.0	253

Percentage distribution of laying fowl by systems of management

System of management	1960/61	1971/72	1972/73	1976/77	1979/84	1988/1992
Free range	30.9	6.1	4.5	2.7	1.9	13.0
Battery	19.3	85.0	88.1	93.2	96.1	85.0
Deep litter and others	49.8	8.9	7.4	4.1	2.0	2.0

Why should poultry keeping have become such a highly intensive operation? Firstly, we may note that the disadvantages of keeping birds outdoors can be considerable and overpowering. Food costs become much heavier because colder conditions require a certain, and often large, percentage of the food merely to keep the birds warm and to provide the energy required for exercise. There may also be severe losses of food when it is gathered by wild birds, rodents and other predators, whilst labour requirements are infinitely more demanding and much greater areas of land are required. These disadvantages represent the main reasons why the poultry industry 'went intensive' in the 1940s and 1950s but since that time there have been further powerful arguments for the development of much fuller environmental control in poultry houses. Firstly, for reasons of improved productivity, the need to control the lighting, both in duration, frequency and intensity, made it highly desirable not only to house poultry but also to keep them under artificial light programmes in windowless houses. Later, still further investigations on the physiological needs of poultry showed the importance of warmer housing as a means of improving productivity and lowering feed costs, whilst the need to fully *control* the feeding rather than just *provide* the food with many types of poultry has further emphasised the desirability of controlling the total environment of birds of all ages. Finally, keeping the birds under control has enabled better disease prevention and the effective administration of protective vaccination and therapy. With all these reasons as background, mammoth units of 100 000 birds or more have been built up all over the world requiring a great expertise in their planning and subsequent management. These large units contain either commercial layers in cages or on 'built-up' (deep) litter, or meat birds (broilers and capons) which are kept almost entirely on deep litter. Deep litter is also used in smaller units concentrating on breeding and rearing, whilst some laying and breeding stock and their replacements are still kept on the traditional extensive or semi-intensive systems.

This, whether one likes it or not, is where we are at the moment and it is a matter of intense dispute whether certain aspects of poultry husbandry are such that the welfare of the birds is in jeopardy. Throughout this book, this aspect will be considered, but it may be appropriate to put the two sides of the controversy in brief outline now. One argument is that an intensive system properly managed with scrupulous attention paid to every detailed aspect is the more humane. Birds may be inspected readily, food and water are also present and the microclimate can be so carefully regulated that there are comfortable and correct conditions at all times. Should disease occur, it can be diagnosed and treated without any delay. There is no risk of predators harming them.

Conversely, however, there are a number of ways in which the management can go wrong, with the ever present risk that large numbers of birds can be cruelly treated if standards deteriorate. There may be mistakes of a serious nature in the design of houses, cages or equipment; or the birds may be housed too densely, or the environment may become unduly hot or cold imposing stress on the birds. Food may be incorrect in composition or seriously deficient in one or more elements. An automatic system of delivering food or water may break down and its malfunction go undetected. A careless attitude or slow response to the prevention or treatment of disease may cause suffering and when large numbers of birds are housed together infections can spread very rapidly. In particular, there should be the greatest emphasis laid on the necessity of having alarm devices to warn of any failure in the electric supply or a breakdown in equipment. There must be either the installation of a stand-by generator or a method for the manual provision of ventilation and other essentials to cope with any period when power is deficient.

So we can see that in modern poultry management a very high standard of knowledge and of responsibility is called for if the welfare, good productivity and health are to be ensured. The aim of this book is to present the information so that the ability to achieve these standards can be attained. One feature that is apparent from Table 1.1 is that we are now seeing a considerable change in the systems, with some 13 per cent of eggs now produced by 'free range' birds.

Breeding

Those domestic fowl producing commercial eggs are divided into two main groups, the light and heavy breeds. The light breeds, and that means predominantly the White Leghorn breed, are small birds 1.5–1.9 kg in weight which lay white shelled eggs. They are economical on food due to their small size and they have the advantage of rarely going broody. In contrast the heavy breeds, examples of which are the Rhode Island Red and Light Sussex, are heavier birds of 1.9–2.2 kg and need more food for their maintenance. They have a tendency to go broody and lay a brown shelled egg which in some countries sells at a higher price than the white shelled egg – though it must be stressed that there is no difference whatsoever in the quality of brown and white shelled eggs. However, the heavier breeds with their bigger carcases, command a substantially better 'salvage' price at the end of their laying season, especially if the birds have been kept in cages, since the absence of activity ensures that the flesh is tender.

The egg farmer rarely uses these pure breeds nowadays, preferring to purchase the chicks or older birds from one of the large breeding concerns which specialise in producing hybrid chickens. Hybrids are genetically complex mixtures of strains and breeds of a light, heavy or a light–heavy mixture, producing eggs with white, brown or tinted shells. The breeding of hybrid poultry involves two basic stages. Firstly, numbers of closed families are built up by inbreeding and selecting for the chosen characteristics. These are then crossed in many complex arrangements in an attempt to produce an offspring which is a layer of the highest quality. The White Leghorn remains the basis for the very prolific layer of white shelled eggs and the Rhode Island Red for the strains producing brown shelled eggs. Tinted egg layers or 'Medium Crosses' may be crosses between White Leghorn and Rhode Island Red or White Leghorn and Light Sussex.

Similar breeding programmes based on 'heavy-weights', such as the White Rock and Cornish breeds have been developed for the production of meat, chiefly broiler chickens, using separate sire and dam strains.

The challenges presented by these specialised breeding operations can be appreciated if consideration is given to the principal characteristics of the birds produced. The qualities of a good layer, for example, include a high yield of large eggs with both contents and shells of good quality;

minimum nutrition for maintenance and production requirements; good liveability; resistance to disease; absence of broodiness; docile temperament and good feathering. The stock from which replacements are bred must produce large numbers of hatchable eggs which will give rise to healthy and vigorous chicks. Broilers must have quick growth, good food conversion, good conformation, correct colour of flesh and feathering; good liveability and resistance to disease. From the breeding point of view the female strain should produce large numbers of eggs with a high hatchability, while the male strain should have a good growth rate, high food conversion efficiency and fine conformation.

The organisations that now produce birds for commercial egg or meat production are totally different concerns from the small breeders who used to supply the market up to about thirty years ago. In order to keep pace with progress in this field only large companies can provide the necessary facilities for continuous research and development. The result has been that only a comparatively few breeding organisations have now been left world-wide. Whilst the responsibility that they bear is immense there is every indication that they can meet the need and the fears sometimes expressed that the genetic pools are becoming too concen-

AN OUTLINE OF THE ORGANISATION REQUIRED FOR MODERN POULTRY PRODUCTION

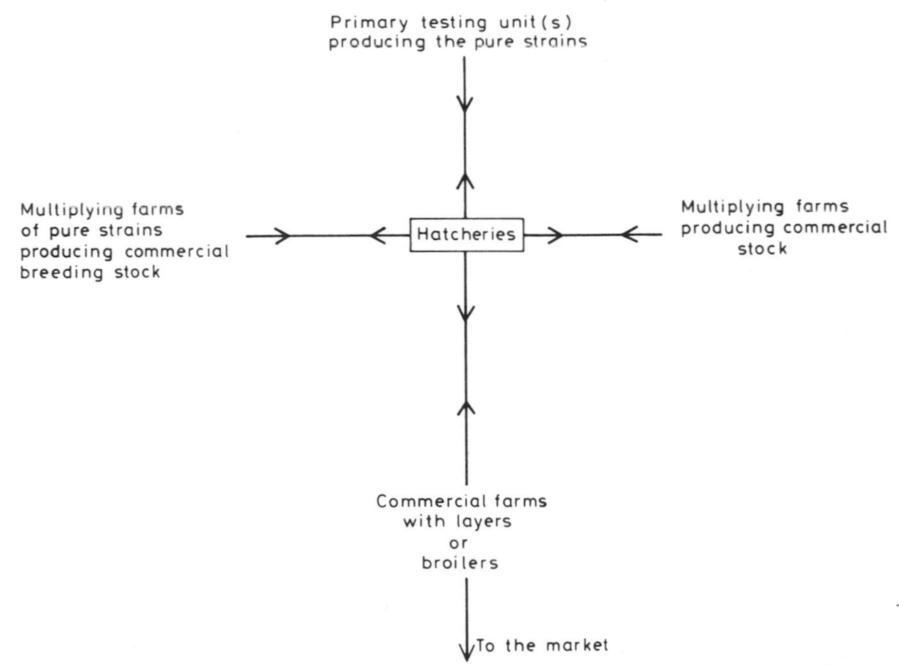

trated do not stand up to practical scrutiny. The fact is that each company in its own right maintains an enormous supply of breeding material and the skills are available to make use of it. It is an intriguing fact that a majority of the large breeding companies are in the hands of industrial conglomerates amongst which food and pharmaceutical concerns predominate. Both should be well aware of the risks taken in undue concentration of their resources.

Plate 1 Three tiers of 'stacked' cages containing five heavy birds ('brown eggers') per cage

Plate 2 Egg collection is automated by a moving belt, the eggs being conveyed to the revolving collecting device as shown. It is the ease of egg collection with cage systems which above all other factors makes them so much more competitive than other systems in management costs in the large commercial unit.

Plate 3 Broiler breeders in a hot climate in naturally ventilated houses but with electric circulating fans mounted under the ridge.

Plate 4 Hatching eggs being placed in the sanitising machine before entry to the incubator.

2 Disease and intensification

In the planning of modern intensive poultry units very little attention has generally been given to the effect of such intensification on disease, its incidence and control. Yet there is now abundant evidence that the productivity of poultry falls as the size of the unit increases and when there is no apparent difference in management between sites. In a survey carried out by the author, broilers showed a variation in finishing weight at the same age from 2.1 kg in groups of 20 to 1.4 kg in groups of 30 000 and an almost pro rata relationship with groups between of 50, 100, 500 and 10 000. These differences occurred with birds from the same genetic material, eating similar food and without any obvious disease. A similar trend was shown in a survey carried out by the Agricultural Development and Advisory Service – broiler growers with a location size of between 595 m^2 and 738 m^2, which is a modest range, gave the best performance. Then, for each doubling of location size beyond this average, weights decreased by 0.09 kg. The decline in weight was almost certainly due to the increased incidence of disease and with such accurate figures available to the poultry industry good use should be made of them in modifying old units or planning new ones. The same experience has been shown with productivity in laying units – above a certain size productivity tends to fall away.

The disease incidence that is apparently greater in the larger units is not necessarily due to obvious 'clinical infections' and the picture is often confused and so unclear that the nature of the problems goes undetected. It is of interest that it has been reported that some animals can produce substances which inhibit the growth of smaller animals of the same species. The amount of substance produced and its effect are proportional to body weight and the result is that the more rapidly-growing animals survive at the expense of the slower-growing ones. This phenomenon could be a factor in the depression of overall weight in large groups and the increasingly uneven growth that occurs, in addition to the problems of disease and behavioural stress.

Husbandry, size and efficiency

In relation to the facts given in the preceding paragraph, it is one of the encouraging elements in poultry husbandry that, in spite of mass production methods and reliance on automation, birds are still highly responsive to the effects of management. No more and no less than other farm livestock, birds with the same apparent housing, nutrition, facilities and of the same genetic material are still capable of giving *vastly* different results depending on the care taken with management. It is significant that usually the farms which head the tables of productivity are the smaller units in the personal charge of the owner. These are more efficient on several grounds as a way of producing *any* poultry products.

If productivity is better and food conversion improved then clearly the world's resources are being used most economically. Thus the very large unit may seem profitable as a purely financial exercise because of the very large numbers of stock, but as a means of utilising the world's limited resources this may not be so at all and if productivity, food conversion, mortality are worse and disease incidence is higher, then it can be a bad bargain and may also be producing an inferior end-product. With a smaller unit it is also easier and indeed more likely that better use will be made of the 'waste products', such as manure and litter, because the land near by may receive it and storage and handling can be kept to a minimum. In future, if the pressure on the need for economy grows as it should, then the highly efficient units must not be so large that they are incapable of giving the best biological results, and are unable to dispose of their waste products easily and usefully.

'All-in, all-out' poultry management

Soon after poultry farming entered the intensive era it was established that there were considerable advantages in following an 'all-in, all-out' programme. This means that all the birds in a unit are housed at the same period, this period being as 'tight' a one as possible, and are then taken through to the end of the growing or productive period so that the whole unit or site can be cleared of livestock, of all muck, both within and without the houses, and then subsequently disinfected, fumigated and rested for at least a day or two. Such a process has an enormous amount to commend it and the list of advantages given below is impressive. It may be pertinent to mention that in very recent times it has become less common to insist on the 'all-in, all-out' procedure and rely rather more on vaccines, antibiotics and chemotherapeutic drugs as a means of controlling disease. All such means have a part to play but it is very doubtful if a system can ultimately survive if it is utterly dependent on them; in any event the two types of measures can go together as aids to good husbandry as they are perfectly complementary.

Here is a list of the advantages:

1. The clearing of all living animals off the site can be the biggest factor in eliminating most, if not all, organisms capable of causing disease. The more frequently this 'depopulation' is practised, the more likely it is that any build-up of disease will be prevented.
2. Once the site has been completely cleared of birds it is possible to apply the most rigorous and effective programmes of disinfection and fumigation, so progressing towards an effective elimination of bacterial, viral, fungal and parasitic infection.
3. The maintenance of birds on a site within a close range makes for a more uniform state of immunity to disease. Mixed ages lead to an immunological confusion so that the uptake of vaccines or treatment are less satisfactory and it is certainly much more difficult for the correct administration of disease prevention programmes as some birds will almost certainly not respond and, or, will receive the treatment at the wrong time.
4. There are sound and practical husbandry advantages. If a site is filled at one time there is only one unavoidable disturbance at this time and at no other. Nothing more will be altered until the site is cleared. If units have birds coming and going at irregular intervals the disturbance occurs at various times and can lead to hazards on health and behavioural grounds.
5. It may also be considered an advantage if the operator can have a pause in the exacting task of management. Maintenance of equipment can also be properly carried out between batches so that there are likely to be far fewer breakdowns.
6. In recent years, with the great increase in intensification, there have been serious problems around poultry sites that are in continuous use, of nuisance caused by flies breeding in the manure and other organic matter. Large sites often have poor arrangements for muck disposal. Smells can cause highly objectionable conditions for nearby residents. An 'all-in, all-out' site can limit the feasible size of a unit, which is not a bad point in itself, but above all it can make it much more practicable to eliminate breeding areas for insects and rodents.

It may be emphasised that the merits of depopulation are in inverse proportion to the age of the birds: it is less important with the adult because by the time a bird has reached maturity it may have achieved a satisfactory immunity to most diseases.

Environmental control and health

Probably the first group of diseases that comes to mind when considering the relationship between environment and disease is the respiratory complex. It is inevitable that under intensive conditions the likelihood of birds being infected with respiratory ailments is greater than those kept extensively or even semi-intensively. Environmental control has certain major functions. Firstly, it must provide ventilation arrangements that constantly bring in fresh air and draw off stale air, gases and other pollutants. In this way, respiratory by-products that may contain certain disease-producing organisms will be removed. It must, however, perform this vital function in such a way that a uniform movement of the air is applied to the whole area inhabited by the birds, so exchanging the old air with the new in a manner that is virtually imperceptible to the stock.

In a later section of this book (chapter 7) details are given of the practical way in which good ventilation can be achieved but the reasoning behind the systems should be understood.

As birds are placed in ever greater concentrations, it becomes increasingly difficult to get uniform air movement. Far too little attention is given to this and relatively too much to the capacity of the fans, important though the latter may be. It is clear that birds housed intensively do react quite markedly to comparatively minor variations in temperature and air movement and several studies on the performance of birds in cages have shown that there is a significant difference in various locations, with particular emphasis on the deleterious effects of cold and draughts. At the other end of the scale, I have frequently been distressed by heavy mortalities occurring in mechanically ventilated poultry houses after sudden rises in the ambient temperature and humidity. To some extent these may have been due to the higher stocking densities that have been currently practised but more frequently they seem to have arisen from poor siting and design of the fresh air inlets. The mortalities have occurred when the fan capacities have been even more than adequate.

Birds' reaction to illness

When birds become ill with elevated body temperatures and reduced appetites and water consumption, they tend to feel the cold more, just as the human subject does when sick. They will then try to move together huddling in areas of low air movement. Just how much this aggravates the problem it is of course impossible to say precisely but it can be surmised that the build-up and intensification of the disease-producing organism is inevitable under the circumstances, causing a vicious circle of yet more disease challenge to the birds. In the case of accommodation where heat is available, a practical procedure is to increase the heat input when the birds are ailing to raise the temperature by several degrees, thus not only tending to persuade the birds to separate one from another but also ensuring that the ventilation rate is not merely maintained but is increased to support any treatment being used. In emphasising the importance of keeping warmth up to counteract infection, it is also pertinent to stress that many viruses have optimal multiplication temperatures a degree or two below the body temperature of the host so that chilling will promote the harmful effects of these pathogenic organisms.

Control of air movement

It has been stressed that at all times a uniform diffusion of the air across the house is essential. There are two quite different ways of achieving this. The most common is to introduce the air into the building at a low speed and either deflect it away with a baffle or diffuse it through slats, perforated inner linings such as slatted hardboard or media such as mineral, glass wool or canvas. These are frequently used to advantage in 'rescue operations' to diffuse the air into houses with badly designed inlets or to reduce the unfavourable effects of high winds but they have dangers in that they may reduce the total air entry by restricting the efficiency of fans, or, if they are neglected they may become clogged with dust. Thus they require expert installation and use and are considered fully in chapter 7.

The second approach is to bring the air in at a high velocity and either deflect it abruptly against a baffle board close to the inlet or direct the air well away from, usually above, the birds so that the draught potential has quite disappeared by the time the air reaches the birds. It is impossible to state categorically which is the best system and indeed from results in the field it seems that each system, properly designed, can produce equally favourable results. Each form of house seems to have a system most appropriate to its design, and use should be made of the right one in assisting to maintain the health of birds. So many houses have a mixture of different systems that no logical pattern can be said to exist and the essential point to grasp is that most failures occur because they are ill-conceived mixtures of systems.

Another major practical factor in maintaining uniform conditions is the way in which the fans are controlled; there is no need for birds to be subjected to the stress of badly controlled fans. Fans with speed controls enable gentle changes to be made in fan speeds as the climatic conditions change. Most modern arrangements have regulating systems that *gradually* change the speeds of the fans and a variable minimum rate can be changed as required according to the age and the stocking rate of the birds. This arrangement is economic where there are large houses requiring a number of fans but a less sophisticated arrangement may be used satisfactorily with a speed regulator reducing the fan speed to 10 per cent of the maximum. This latter requirement should be specified by the farmer as it is now available on most fans at no extra cost. A unit of two fans, for example, one thermostatically controlled and the other on manual regulation, is as good as the most sophisticated system. Indeed it may have certain advantages because it leaves some important functions to the stockman and does not leave everything to an unfeeling automatic thermostat or thermistor! Recently this type of approach has been more positively advised in multi-fan systems since it requires very simple equipment with little chance of error.

In widespread experience there have been clear indications showing the striking advantages of maintaining laying birds in a warm environment of not less than 21°C (70°F) or more than about 30°C (86°F) with relative humidities between 50 and 60 per cent. If such high temperatures are to be maintained in laying houses all the year round – and in any case they are required during the early rearing stages of chicks – it places an increased emphasis on the importance of maintaining uniform and constant ventilation and the highest standard of insulation. Such high temperatures require especial care in having all the ventilation under control.

Brooding

The detailed practicalities of this will be dealt with in an ensuing chapter but it is pertinent to mention in this section some of the vital features with respect to health. A great deal of work has been done on methods of brooding – that is, on methods and techniques providing warmth in the early days of life – and in some years of work at Cambridge it was found that the best results were obtained from heating systems that combined radiant and convected heat. In practical terms this means an arrangement that provides a warm area for resting birds and a slightly cooler area for movement. Since, however, birds brooded on the floor must be encouraged to use as much of the floor area as possible, right from the beginning, the movement area must still be evenly warmed and a minimum temperature of 21°C (70°F) is advised. In practice, to ensure that the temperature never drops below this level, it is much better to specify a minimum somewhat higher than this, say, 24°C (75°F). Also it is essential that the chicks can be easily inspected and that there is a good uninterrupted air circulation all round them. For these reasons, brooders of largely radiant heat output that can be placed well above the birds are favoured. The heavy insulated canopied type, close to the floor, is no longer necessary in modern buildings that are well insulated overall, and they also have the disadvantage that they concentrate the chicks in too limited an area, and make proper inspection much more difficult.

An alternative arrangement that has achieved some popularity is the use of one heater in each building – a hot-air blower that raises the whole house temperature to the 'required level'. However, the determination of the 'required level' raises some problems. On the one hand, if the health and appetite of the birds are to be ensured the temperature must not be too high, but as opposed to this, hot air requires an essential velocity to distribute the heat so that the cooling effect on the bird is greater than with a radiant brooder. In this connection there are two essentials. Firstly, there must be an arrangement to duct the air efficiently round the house so that it does not have to be blown long distances creating draughts, dust and unaccepable regions. Secondly, whilst the initial temperature must be high, about 31°C (88°F), the method of compensating for the disadvantage for an overall high figure is to reduce this as soon and as regularly as possible. For example, if the temperature starts at 31°C (88°F) and is reduced by 0.6°C (1°F) daily, it will promote the qualities required in the birds. An additional problem with whole house heating is the effect on humidity. With a high temperature over the whole house, the relative humidity will be very low – about 30–40 per cent. It is known, however, that a higher relative humidity, up to about 60 per cent, is needed for the chicks. This promotes good feathering and growth and more particularly is likely to provide the best environment to enable the birds to resist the effects of any respiratory infection. In moister atmospheres the organisms causing respiratory challenge will perish more quickly and the mucous membranes that line the respiratory passages of the birds will remain in a healthier state. Nevertheless it appears that at both *very high* and *very low* humidities, air-borne organisms that can cause disease are likely to remain viable for longer periods than where the humidity is in an intermediate range, that is between 40 and 60 per cent. High humidities, near saturation, such as are found in the winter months, are especially favourable to the survival of respiratory organisms and this may be one of the factors contributing to the much higher incidence of respiratory disease in cold weather. A major reason for the popularity of the radiant gas heating system is the favourable effect it has on the humidity which it keeps at the correct figure.

Ventilation rates

Ventilation rates that are provided for in a controlled environment must be considered with great care. In general, many intensive units would benefit from extra ventilation. Maximum rates have had to be raised in recent years to help to counteract and perhaps eliminate the overall challenging effects of disease in large intensive units. For layers in cages the maximum of 6 m^3/hour/kg body wt (1.5 ft^3/min/lb body wt) is recommended. Note that the ventilation is based on the bird and its weight rather than the house itself, and the rather old-fashioned concept of describing ventilation in terms of 'air change per hour' is no longer an arrangement that is tenable as it makes no allowance for the birds housed. With heavy broiler breeders benefits have been found from even higher rates and levels – up to 11 m^3/hour/kg body wt (2.5 ft^3/min/lb body wt) were the most helpful in the warmest summer weather. The advantages of higher rates for breeders are reflected in better fertility and hatchability which is probably the result of improved air circulation around the body producing a cooling effect and better health.

There is, however, a second very good reason why broiler farmers should re-examine ventilation rates. There have been spectacular improvements in weight recently and this has meant up to 20 per cent more liveweight at the finishing age. Under these circumstances unless fewer birds are put into the building or the killing age is reduced there may be too little ventilation. Insufficient ventilation may exert its effect not only through rising temperature and humidity but also by gaseous contamination. The main gases that can affect poultry and which are likely to be in excessive amounts in the atmosphere in poultry houses are ammonia, carbon dioxide and carbon monoxide. Damage to the respiratory tract, which cannot always be detected if the concentration of the irritating gas is relatively low, may become apparent when the subject is exposed to a challenge with an air-borne infectious micro-organism.

An alternative approach to calculating ventilation rates is to base them on the feed consumption of the birds since this relates to the metabolic rate of the bird and the maximum requirements on this system are 25 m^3/second/tonne of feed per day, and the minimum is 3 m^3/second/tonne of feed per day. Scientifically this is a very sound principle but since in practice it is a rather more complicated way of stating the need, it has not yet found great favour. Calculations based on the two alternatives do not differ very much.

Further details on the practical methods of ventilation are given in chapter 7.

Vaccination and the environment

The points just made about health and environment have a special relevance in relation to the use of vaccination against respiratory infections such as Newcastle disease, infectious bronchitis and infectious laryngo-tracheitis, especially when given in the highly effective form of a droplet spray. These are live vaccines, which means that they are capable of giving a good, and to some extent an immediate, protection but they are, by their very nature, mild forms of infection, so that the side effects can only be eliminated if the birds are kept in the best environment.

Litter management

The management of the 'deep' and 'built-up' litter in a poultry house is of the greatest importance and in truth seems to be one of the most neglected aspects of poultry husbandry. It is frightening to see broilers, layers and breeders maintained virtually throughout the winter months on accumulations of their own droppings and in not a few cases it is possible to have one's boots sucked off in the quagmire – it has happened to me! Birds cannot possibly thrive under such conditions. Parasitic and bacterial infections are highly likely and the poultryman's enthusiasm for the job falls away as he sees the hopelessness of the situation, as no action he can take will materially improve the conditions. Probably the most serious consequences of all are in breeder houses where wet litter can have a calamitous effect on the feet of the cocks, causing accumulations of infected litter on the feet subsequently leading to a fall in the level of fertility.

Good litter needs care – it is not achieved by accident. A start must be made with adequate material which can be soft wood shavings, or chopped straw, or a mixture of these. Some poultry farmers are using shredded paper; 'old' litter from previous crops may be used with advantage provided it has been completely stacked and heated. A depth of 150 mm (6 inches) at least is required and it should be placed on a dry damp-free base. Studies have shown that litter on an earth base rather than on a damp-proofed concrete floor will contain as much as 10 per cent more moisture on average, so that under these circumstances it may be more difficult, though by no means impossible, to manage. Equally important is the construction of the walls of the house. With an insulated house the insulation must be maintained to the ground – which it rarely is as the dwarf wall taking the prefabricated base is usually devoid of it – and no damp penetration must take place. In any case, because the edges of the floor are usually the coldest spots this is often where the litter starts caking. The consequence of caked litter at the edge of the floor, with birds congregating in this area, is to compound the whole problem often with disastrous results.

The ease with which the litter is maintained in a friable state is greatly influenced by the environmental conditions in the house, all of which are explained in detail in chapters 5–7. Above all, uniform temperatures and air movement are essential to good litter conditions and an even distribution of birds over the floor and modern systems of distribution of the air by diffusion of incoming air are capable of giving the best results.

'Danger' areas in litter management are drinker points, due to splashing, and feeding areas, due to concentration of birds. It may be essential to turn the litter quite frequently and it is often desirable to turn it all from time to time and especially before it is working properly. There is no denying that this is a very laborious task but it can be aided by various mechanical implements which help enormously. The important thing to appreciate is that once the litter is 'working' the activity of the birds themselves will keep most, if not all of it, in a good condition. The activity benefits the birds, they obtain some nutrients from the litter and the whole atmosphere and environment in the house can be pleasant, not to mention the eventual manurial value of the litter. Working litter is warm and adds warmth to the house but wet litter is colder and takes heat from the house in an attempt to dry out.

Whilst it is usual to place about 150 mm depth of shavings or other material on the floor to start with, it is in fact even better to place a depth of about 70 mm to start with and then add to it as necessary. The smaller the depth of litter that is started the more likely it is to be totally 'worked' by the birds and indeed if a great depth is put in at once it is very possible that the bottom part is never moved at all through the cycle – especially with broilers – so that its presence is entirely useless and wasteful. By adding litter later, droppings are diluted, the activity of the birds is enhanced, the condition of the litter is improved, and full use of all areas is helped. Those who have been in the poultry industry some years will recall that when deep litter was first used it was always put down in the summer months and often incubated with some horse manure or other well established manure to ensure immediate activity.

When litter is heaped, heated and re-used it is inadvisable to place it in the highly-heated brooder area. Firstly, it may present rather too massive a challenge of potentially pathogenic organisms. However, more important is the danger that can result under these circumstances of high levels of ammonia from the warmed litter.

High ammonia levels for all ages of poultry are potentially dangerous, and they are also most unpleasant for the operator. Levels of up to 15–20 parts per million (p.p.m.) are acceptable. If levels go over 40 p.p.m. there may be some reduction in food intake but if levels go over 50 p.p.m., the delicate membranes lining the respiratory tract are affected and respiratory disease is much more likely and even blindness can result. By and large it is possible to estimate the levels of ammonia fairly accurately by using one's sense of smell – if it is definitely 'in the air' then it is really too high – but there are more accurate ways of getting an estimate, either by using a litmus paper colour indicator (marketed by Vineland Laboratories), or more accurately by using a 'Dräger' gas detector, which can detect levels of a large number of gases by pumping gas, by means of hand bellows, through indicator tubes, which enable an immediate reading to be obtained.

The welfare of poultry

Highly intensive methods of poultry husbandry have resulted in the expression of a considerable measure of public disquiet as to the humanity of the methods used. The objections have come under two principal headings, firstly, whether we unduly restrict the movements by putting them in cages or giving them little room for movement in other systems, and secondly, whether we deny them, under some systems, their natural instinctive urges to dust bath, stretch their wings, and scratch in litter. A great deal of work is being undertaken by poultry behaviourists in many countries to try and establish the validity or otherwise of these fears and it is likely the questions will be answered in due course. In the meantime it is important that poultry farmers should be aware of the local law pertaining to the Welfare of Poultry. There has been a tendency in most countries, as in European contries including the U.K., to rely on 'codes of practice' which is in many respects a very wise way to regulate methods pending the findings of relevant research and investigation. Many of the codes are based on standards which are capable of achieving good productivity and for this reason alone are very worthy of attention. Full consideration of poultry welfare and the use of alternative arrangements to cages and other restrictive systems is given in chapter 17.

Plate 5 Interior of a large wide-span poultry house for broilers after cleaning and fumigation. Note the litter on the floor, which is usually put in place between disinfection and fumigation procedures so that it can benefit from the formaldehyde gassing. The automatic feeding vertical piping is seen under the ceiling where it is safely placed during disinfection and fumigation. Ventilation is by side-wall fans and controllable ridge intake – a satisfactory system using high-speed air flow at roof level.

Plate 6 A contrasted ventilation arrangement to *Plate 5* in which the air is blown into the roof space with fans in the gables and diffuses into the house at a very low speed through a 50 mm thick glass-fibre ceiling.

3 Poultry nutrition

The principles

In order to understand the principles behind the feeding of the chicken it is helpful to look at the way in which the bird digests its food (Fig. 3.1). The digestive system is simple, short but extremely efficient. The beak gathers the food material and at its edges are tactile cells by which the bird decides whether it will accept the food or reject it. This is decided on the basis of reflectivity and feel and some taste, though the taste buds are few in number. In certain systems of poultry husbandry it may be the practice to pare or trim the ends of the beak to prevent feather-pulling or cannibalism. Such a procedure may impair the correct selection and apprehension of the food which emphasises the need to carry out the 'surgery' with the utmost care, if at all. There is no evidence that chickens have any fine or real ability to smell.

The food is swallowed whole, with a little saliva added, passing down the oesophagus to the crop which is basically a container for bulky foods. Here the fibre is well softened and the food acidified (to about pH 4) by lactic acid produced by bacterial action. From the crop the food passes to the proventriculus, which secretes hydrochloric acid and pepsin, and which is the organ that most closely corresponds to a mammal's stomach. From here the food goes to the gizzard which is a strong muscular organ which contracts powerfully and rhythmically, reducing the contents to a thick, paste-like mass with the aid of the insoluble grit present within it. The food next goes through the various regions of the small intestine by the regular so-called 'peristaltic' contractions and it is in this part of the digestive tract that the digestion and absorption of the food principally takes place. Digestion also occurs to a lesser degree in the caeca which are two blind sacs at the junction of the small and large intestines, the latter being chiefly concerned with the absorption of water. From here the faeces pass for final evacuation to the cloaca, which is also concerned with the excretion of urine, the acceptance or delivery of semen, and the passage of the egg to the outside.

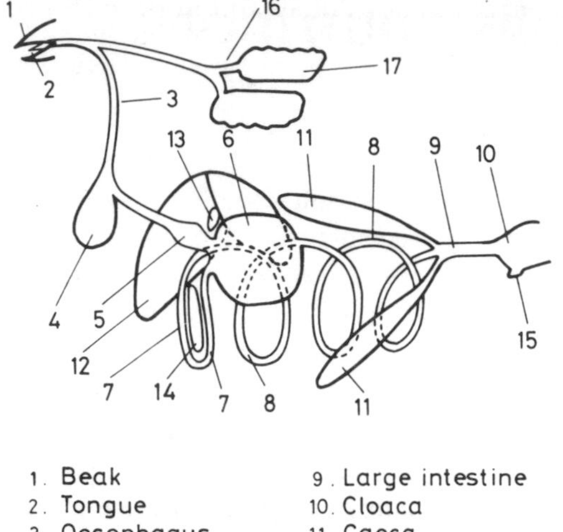

Fig. 3.1 Diagrammatic representation of the digestive and respiratory systems of the domestic fowl.

The condition of the droppings of the fowl is a good guide to its state of heatlth. Wet droppings are indicative of either a nutritional abnormality – for example excessive salt in the ration or rancid oil – or any generalised 'systemic' infection. Green droppings indicate a liver (bile) disorder, white droppings a kidney infection and red ones show the presence of blood which may be due to either coccidiosis or an acute bacterial infection.

The basic feeding essentials

Poultry, which are normally housed without access to soil, or herbage, and generally without any degree of sunlight, must be provided with food which is not merely sufficient for their needs but in which a carefully calculated allowance is made for a safe margin of additional needs. In recent years there have been several cases of the nutritionist being 'caught out' by the very unexpected and high demands for certain elements, often only required in minute quantities. These have arisen more often than not from the successes of the geneticist in producing birds with greatly enhanced productivity in every way, so outstripping the normal nutritional requirements both absolutely and in proportion, the latter really being just as vital.

Protein and energy

The first essentials for the provision of correct, complete and balanced poultry rations are that they should contain the appropriate amount and proportions of energy and protein. Dealing firstly with the energy fraction, this is needed for the maintenance of all normal functions and any amounts supplied over and above these basic needs will be used for production – which in the fowl is for eggs or growth. If there is excessive energy in the feed, then undesirable amounts of fat production and deposition will result which will be harmful to the metabolic processes and, or, the marketable product.

The term now used for the assessment of the energy is 'metabolisable energy' (ME). Briefly the ME content of a feed is that part of it which is truly available (or metabolisable) to the animal, after subtraction of the energy in the droppings. Even some of this metabolisable energy is wasted, for heat is produced during digestion and absorption of the feed. The final amount of energy used in maintenance and production is rather less than the ME and is in the 'net energy' (NE). However, it is considered that the ME value is the most realistic value to use so foods are usually classified and the bird's requirements specified in terms of ME values or needs.

Energy has for many years been measured in calories, a calorie being the amount of heat necessary to raise the temperature of 1 gram of water by 1°C. As the calorie is such a small unit the usual measure is the kilocalorie (kcal) which is equal to 1000 calories. Now, however, the S.I. (international) metric system uses the megajoule (MJ). One megajoule is equal to 240 kcal. The poultry industry in general, at the time of writing this book, still prefers the use, in practice, of the kilocalorie to the megajoule, so that the two units will be quoted in this chapter.

Diets may be classified as high energy (over 2860 kilocalories of ME per kilogram weight of food or 11.9 MJ/ME/kg); medium energy (2640–2860 kcal/ME/kg or 11.0–11.9 MJ/ME/kg); or low energy (2530–2640 kcal/ME/kg or 10.5–11.0 MJ/ME/kg). High energy diets may be used for small hybrid laying pullets, referred to previously, so that they may produce the very large number of eggs in relation to their size. Though they may weigh only a little over 1.5 kg (3 lb) they may consume less than 110 g (4 oz) of food daily. Broilers need high energy diets too as they must reach about 2 kg (4.4 lb) in under eight weeks at a food conversion of about 2, so consuming only about 4 kg (8.8 lb) of food to achieve what are by any standards remarkable weight gains. By contrast, for heavy hybrid and rearing diets, the same high energy diets would be not only wasteful but they would be actually harmful, since the birds would become excessively fat. In particular, the supremely heavy weight broiler breeding stock must be checked from laying too early in their development and should receive an appropriately lower energy diet to ensure this. In the high energy diets the main cereals are maize and wheat, but where a low energy diet is required the proportion of wheat by-products and barley and oats can be raised in relation to the maize and wheat to reduce the overall ME figure. Where extremely rapid growth and weight gains are needed, as in broilers, the energy level can be raised by including in the ration up to about 5 per cent of added fat.

Protein is required for the general body development in all growing birds, and layers also need a good proportion in their rations since an egg consists of 13–14 per cent protein. The young chick requires 19–23 per cent of a good quality protein in its diet for the first six weeks but thereafter the level can be reduced to 16–18 per cent and in the case of broiler breeding stock, where fast growth is undesirable, to 14–16 per cent. It is often good practice with heavy type rearers to feed a supplement of grain with the concentrate ration but whenever this is done it is necessary to adjust the levels of protein, minerals and vitamins upwards in the remainder of the ration in order to retain a fully balanced mixture of food – the grain being of course relatively deficient in elements other than carbohydrates. The same precaution is necessary for birds reared on pasture which are usually fed enough concentrate and grain to provide over half the diet, the remainder being obtainable from the pasture if it is of good quality and especially during the earlier part of the growing season. It is vitally important to understand that birds will eat sufficient feed to satisfy their *energy* requirements and this is one of the main factors limiting intake. Hence the consumption of a high energy food will be lighter than that of a low energy one and the protein, vitamin and mineral content must be adjusted accordingly in the former type of rations.

The quality of the protein is also another essential element which is rarely given the consideration it should. The two most important amino acids, which form the constituent parts of the protein, are for chickens lysine and methionine, and the proteins richest in these are those derived from animal origin. For this reason animal protein is almost always included in chick and broiler diets, though it it is not essential for laying birds on medium or low energy diets. In some diets synthetic amino acids are now being used as an economic method of adding these to improve the quality of the protein. The best quality protein of all for poultry is white fishmeal, which may often be included in the rations of young chicks at levels of up to 10 per cent, this amount being reduced as the birds get older. Meat meal products are also used but often contain forms of protein inferior to those in white fishmeal. The best vegetable protein by far is soya-bean meal, though it is relatively poor in methionine and should be fed to chicks as part of and *not* the sole source of protein. Sunflower meal on the other hand is rich in methionine and may be fed sparingly to chicks although its high fibre content limits the quantities that can be included in the ration.

Vitamins

Housed poultry are entirely dependent on the vitamins that are present in their compounded food in the correct amount and proportions and any interruption in their supply can give rise to most serious consequences, sometimes with striking rapidity. As well as naturally-occurring sources of vitamins an increasing number are provided in synthetic form. The principal vitamin requirements are as follows.

Vitamin A The principal feature of vitamin A is its function in ensuring adequate growth and as a means of assisting in the birds' resistance to disease. It may be formed by synthesis in the body of the bird from carotene present in green vegetable matter or yellow maize, or it may be fed direct either by fortifying the ration with stabilised cod or halibut liver oil, or alternatively by using Vitamin A in synthetic form. Chick diets should contain approximately 12 m.i.u. (million international units) per tonne of feed, laying diets 6 m.i.u./tonne, and diets for breeding stock between 10 and 12.

Vitamin D Chickens can synthesise this vitamin from sunlight, but even if they are kept under natural extensive or semi-intensive systems, the amount is totally unreliable and usually insufficient. Like vitamin A, it may be supplied either in fish liver oil, or in synthetic form. It should be given to poultry as vitamin D_3 and not D_2, as the latter is poorly utilised by poultry. Unless vitamin D_3 is present in sufficient amounts, the bird is unable to utilise properly the calcium and phosphorus in the diet which are vital for bone formation and for the production of a good quality egg shell. The requirement for vitamin D_3 in the diet is as follows: 3–4 m.i.u./ tonne of feed for chicks, and 3 m.i.u./tonne for layers and breeders.

Vitamin E A deficiency of this vitamin causes a disease of the nervous system in chicks known as 'crazy chick disease'. It is also essential to breeding stock for the good hatchability of their eggs. Whilst there ought to be a sufficient supply in the cereals of the diet this cannot be guaranteed and in any case vitamin E is very readily destroyed by bad storage, or over-heating, for example if food is placed under brooders, or by rancid oil or fats in the food. Poultry foods should contain up to 10 mg/kg of added vitamin E but there is considerable variation between the needs of different classes of birds with the highest amounts required for breeders and young, fast-growing birds. This is one of the most crucial of vitamins and it appears, in practice, that it is remarkably easily destroyed between the time it is added to the ration and the time it is put in the *trough* emphasising the need for expert handling of all compounds and their ingredients at every stage.

Vitamin K Vitamin K is an essential element in the mechanism of the clotting of the blood and a deficiency can lead to multiple haemorrhage. It is present naturally in all green foods; alternatively, a small quantity of a product which is rich in it, such as lucerne meal, can be added to the ration, or a synthetic form can be added to compound rations. The needs are small, and 2 i.u./kg will suffice under normal conditions. It has, however, been widely used by poultrymen as a therapeutic additive – usually in the drinking water – to assist in a cure when birds are affected by conditions causing haemorrhaging, such as gangrenous dermatitis (page 133) or haemorrhagic disease.

The B vitamins These are well distributed in cereals and deficiencies are normally unlikely, except in the case of riboflavin (vitamin B_2), but the vitamins are frequently added in synthetic form to ensure a constant amount and a safe margin. The main functions of the B vitamins are as follows. All the B vitamins assist the chick in achieving its optimum growth. Vitamin B_1 (also known as aneurin and thiamin) is needed for the metabolism of carbohydrates, while a deficiency of vitamin B_2 (riboflavin) causes a condition known as 'curled toe paralysis' in chicks. Another one, nicotinic acid is required especially for proper feathering and choline is associated with the metabolism of fat. Pantothenic acid deficiency leads to dermatitis. It is also needed for normal growth and for good hatchability and the proper feathering of the bird. Vitamin B_{12} (cobalmin) is required for the development of normal red blood cells and also for good hatchability. It is especially important for young birds and breeders. Folic acid shortage can lead to anaemia and weaknesses of the legs.

Substantial quantities of B vitamins are found in all the following foods: grains, seeds, fish meal, liver meal, meat meal, soya-bean meal, dried grain, dried milk, dried yeast, sunflower meal, unheated offals. However, it should be stressed that vitamin B_{12} is found only in foods of animal origin, or it can be synthesised by bacteria from dung or deep litter. The levels required by chicks (mg per kg of food) for the major B vitamins are as follows: thiamin 1.0, riboflavin 5.0, nicotinic acid 28.0, pantothenic acid 10.0, folic acid 1.5, choline 1300, B_{12} 0.02.

A further important note is necessary relating to the B vitamins. In recent years there have been serious outbreaks from time to time in chicks, and especially broilers, of a condition known as 'fatty liver kidney syndrome' (FLKS) which has been found to arise from insufficient biotin, a B vitamin, in the ration, especially where wheat or barley constitute the majority of the cereal fractions and if the protein element does not contain substantial quantities of soya-bean, groundnut or sunflower meal. The total allowance of biotin should be up to 0.15 mg/kg of diet in broilers and 0.18 in broiler breeders. Actual supplementation of biotin over and above the levels normally likely to be present in the food would be 0.08 and 0.12 respectively. A major factor in the inducement of the FLKS is when the energy/protein ratio is widened. Peak mortality has been found to occur in diets with an ME of 3000 kcal/kg and crude protein of 18 per cent, a ratio of 166:1 which is too high for the chick (see also chapters 4 and 11).

Minerals

Calcium and phosphorus The main function of these two minerals is in the make-up of the bones of the body. It is not only that these are required in sufficient quantity but also in the correct proportions. For the young chick the ratio should be calcium to phosphorus 2:1 to 1:1 and not outside this range. There is some calcium in all foods but those of animal origin are generally much richer in it than are those derived from vegetable sources. Phosphorus is also present in all foods, but in cereals it may be in a form which is poorly absorbed so that supplements may well be needed. Furthermore, the proper absorptions of calcium and phosphorus are possibly only if sufficient vitamin D_3 and manganese are present. The young chick needs a minimum of 1 per cent of the diet as calcium and 0.5 per cent as available phosphorus, whilst the laying hen needs about 3.5 per cent of calcium (approximately 4.0 g daily) since this is the main constituent of the egg shell. Calcium and phosphorus should be added in the correct quantity and ratio either as steamed bone flour or as dicalcium phosphate, unless the diet contains at least 5 per cent of fish or meat and bone meal. There has been recent evidence that special care must be taken to ensure that sufficient available phosphorus is given and the level may with advantage be increased considerably. In field trials with broilers, levels up to 0.7 per cent have helped to alleviate the ever present problem of malformation of the skeleton and the poor calcification of cartilage. Expert advice should be sought before indulging in high levels which almost constitute therapeutic measures.

Manganese This mineral forms a link in the chain of the calcium metabolism of the bird and is required by breeding birds to ensure good hatchability of their eggs, by layers to give good shell strength and by all growers to promote bone formation. A very characteristic effect of its deficiency is the condition known as perosis when the Achilles tendon slips off its groove behind the hock joint, pulling sideways and backwards. However, a deficiency of choline can also give rise to perosis and to abnormalities in the metabolism of calcium and phosphorus. Many of the products fed to poultry, such as wheat and limestone, should contain sufficient manganese; if a supplement is required it is customarily given in the form of manganese sulphate.

Iron, copper and cobalt These elements are essential for the formation of haemoglobin and the ration may be strengthened with them to ensure an adequate supply.

Iodine. This is normally present in the usual foods, particularly fish meals, but since a deficiency can cause poor hatchability, compounders often prefer to add it as a safeguard.

Common salt This is essential for protein digestion and is normally added to rations at amounts of up to 0.5 per cent. Any excess of salt which will result in thirst and wet droppings must be avoided. There will also be serious deleterious effects on productivity, whether it be growth or egg production either from an excess or a deficiency of salt.

Water Poultry have very large demands for water and it is absolutely essential that it is available at all times in all systems of poultry management. Not only will production suffer markedly if birds have insufficient to drink, but in the young chick and in birds of any age in hot weather, it may take only a few hours for death to take place if there is total absence of water. There are many ways in which birds can be deprived of water accidentally. Automatic drinkers may run dry unnoticed and this is especially liable to happen with nipple drinkers that have no reservoir where the water can be seen. Sometimes, as well, the positioning or height of the drinker is such that some birds cannot reach the supply properly.

Daily water consumption in litres per 100 birds:

Age in weeks	Water consumption per 100 birds in litres
1 day – 2	3.4–4.5
2– 6	7–10
6–10	14
10–20	18

A relatively new practice is to control the water consumption of growing breeders on restricted diets. This is because birds that are restricted in their feed tend to drink too much water. In programmes of water restriction it is usually provided for periods of about 4 to 5 hours daily but the procedure advised by the producers of the birds should be followed precisely. Great care must also be taken not to overdo the restriction and very serious consequences can result if the temperature rises above 30°C (85°F) and there is insufficient water available to allow the birds to pant and evaporate off their heat.

4 Practical poultry feeding

There are several forms in which a balanced ration may be fed to poultry – all dry mash, a mixture of dry mash and grain, wet mash, crumbs or pellets plus various combinations of these. All dry mash is especially useful to the farmer who makes up his own rations and because it takes longer to consume it can help to prevent vices such as feather pecking and cannibalism. Wet mash feeding is little used nowadays because of the labour involved (it must be made up freshly each time), but birds find it very palatable, and it is eaten more quickly and with less wastage than all dry mash. A mixture of grain balanced with mash is very suitable for birds on deep litter or built-up straw because the retrieval of the grain scattered on the floor exercises the birds, it helps to allay vices and it tends to create better conditions in the litter. There appears, surprisingly, to be little wastage if the general state of the litter is good.

Pellets and crumbs may be favoured in modern poultry feeding systems as there is a minimum of wastage, and since each pellet or crumb is a completely balanced ration in itself there is no opportunity for the birds to practice selective feeding. Nevertheless, because of the advantages of dry meal just referred to, it is likely to be the foodstuff of choice with birds in cages.

Restriction of diets for poultry

In general birds, and indeed all animals, tend to eat rather more than they need, especially if they are placed in an environment where there is little else to do but eat and drink. In a sense over-eating is a tendency to a vice as much as feather-pecking or vent-pecking. Apart from the economic undesirability of this, it may also make the birds too fat and, or, unable to produce as well or efficiently as they should. If it could be arranged it would probably be desirable after an age of 4–5 weeks for some form of restriction to be used which would be beneficial both for the birds and for food conversion efficiency. Methods of restriction are currently being investigated very carefully and there is still conflicting evidence both on their value and the best technique. In rearing it seems most likely that a restriction of the *quantity* of the food actually consumed is the best way. Other methods exist that are apparently not so good. For example, the time for which the birds have access to food can be limited but the birds may learn to consume larger quantities in a shorter period, so effectively defeating the technique! An alternative is to lower the energy value of the food but this may not be economical because the birds over-feed to compensate, so far as they can, for the lower energy. Yet another procedure reduces the protein and, or, the amino acid content of the food, the particular amino acid being lowered is generally lysine. While this method can be successful it may lead to birds consuming more food merely to correct the imbalance deliberately engineered. It appears, therefore, that restriction of a properly balanced diet by the administration of known quantities of it is the best technique.

The prospect of food saving in the laying birds is even greater but the risks are inherently greater too if the wrong method is used, or the wrong quantities or birds are involved. The effects are potentially good; egg numbers should not be reduced but weight may be marginally less: food conversion in particular is improved but there will be some reduction in body weight. The best saving should be with heavier strains, the least with the lightest, where it may not be a desirable process at all.

There are several practical techniques of dealing with food restriction in the hen. Probably the best is to give measured quantities once a day, and preferably in the afternoon, which reduces the energy intake by some 8 per cent. The reason why it should be given in the afternoon is interesting in that it has been found that the peak requirement for nutrients is late in the day, especially for protein, energy and calcium. Another way of restriction is to limit the feed to two 2-hourly periods, morning and afternoon, which is likely to have a similar overall effect. A rather simple technique is to let the birds have an *ad lib* diet up to an age of about 36 weeks and then allow them no more daily than their appetite demanded then. A very effective form of restriction, used for many years with broilers, is to give them alternating light and dark periods, increasing the proportion of dark periods as they grow. More details are given on this in the section on lighting in chapter 5 and in the section on breeder management in chapter 10.

The construction of a poultry ration

The complete compounded poultry ration is more complex than almost any other diet used on the farm and few poultry farmers mix their own

Table 4.1 Nutritional standards for principal poultry feeds

	Crude protein, %	ME, kcal/kg	ME, MJ/kg	Lysine, g/kg	Methionine + cystine, g/kg	Calcium, %	Phosphorus, %	Salt, %	Zinc, mg/kg	Sodium, mg/kg	Manganese, mg/kg	Vit.A, m.i.u./tonne	Vit.D_3, m.i.u./tonne	Choline, mg/kg
Starting chicks	20	2800	11.66	11	7.5	1.0	0.5	0.40	60	1500	100	12	3–4	1300
Early growers	15	2700	11.25	8	6.0	1.0	0.4	0.40	60	1500	100	12	3	1300
Late growers	12	2700	11.25	6	4.5	1.0	0.4	0.40	50	1500	100	8	3	1300
Light hybrid layers	16–19	2800	11.66	8	4.6	3.6	0.5	0.40	50	1500	100	6	3	600
Medium hybrid layers	15–18	2800	11.66	8	4.6	3.6	0.5	0.40	50	1500	100	6	3	600
Breeders	16	2800	11.66	8	4.8	3.6	0.5	0.40	50	1500	100	10–12	3	1100
Broiler starters	23	3080	12.88	12.5	9.2	1.2	0.5	0.40	50	1500	100	12	4	1300
Broiler finishers	19	3100	12.92	10	7.3	1.0	0.5	0.40	50	1500	100	12	4	1300
Turkey starters	28	2860	11.96	18	8.0	1.0	0.6	0.45	60	1750	120	16	4	1700
Turkey breeders	16	2800	11.66	8	4.0	3.0	0.5	0.45	50	1750	100	16	4	1350

diets because of the considerable problems involved in doing so – and also because the highly competitive state of the feed compound industry ensures the lowest possible cost. Nevertheless, whether or not one does mix one's own poultry diet, it is of great importance at least to be aware of the fundamentals involved in feeding which form the principal

Table 4.2 Composition and nutritive values of some of the commonest foodstuffs used for poultry

Foodstuff	Digestible nutrients, per cent				
	Moisture	Crude protein	Oil (ether extract)	Carbohydrate	Metabolisable energy (kcal/kg)
Barley	14.3	8.8	1.6	54.4	2728
Maize	15.2	8.4	3.2	65.9	3322
Maize germ meal	8.6	15.4	9.8	18.7	2288
Oats	15.2	7.4	4.0	42.1	2376
Sorghum (milo)	13.9	8.4	2.5	61.9	3102
Wheat	15.5	7.9	0.9	64.4	3036
Wheat middlings	13.3	12.7	4.1	35.4	2332
Groundnut meal (decorticated extracted)	13.4	37.8	1.4	26.2	2728
Soya-bean meal	12.7	29.5	15.8	20.8	2552
Feather meal	11.5	87.6	—	—	3102
Fish meal	10.0	54.0	6.2	1.6	2838
Herring meal	9.3	63.7	10.5	2.3	3080
Meat meal	9.8	65.0	11.9	1.8	3740
Milk: dried, skimmed	6.6	27.5	0.8	42.9	2926
Poultry dressing plant by-products meal	9.3	71.5	15.2	0.2	3784
Whey: dried	6.3	10.1	0.6	58.3	2860
Grass: dried	12.9	13.0	4.1	10.7	1342
Lucerne: dried	11.3	12.3	1.1	12.2	1100
Yeast, brewers': dried, unextracted	13.3	37.4	1.6	23.2	2618
Rapeseed meal (extracted)	9.6	34.3	1.7	15.8	1760
Molasses	28.7	4.2	—	37.6	1870

(Adapted with permission from the Nutrient Requirements of Farm Livestock, No. 1. 'Poultry', Summary of Recommendations. (1963) London: Agricultural Research Council.)

Table 4.3 Normal inclusion rates of feedingstuffs in practical poultry diets, per cent

Feedingstuff	Chicks		Pullets		Layers		Breeders (& hens, ducks, turkeys)		Broilers (& ducklings & poults)	
	Min.	Max.	Min.	Max.	Min.	Max.	Min.	Max.	Min.	Max.
Barley meal	0	25	0	40	0	40	0	40	0	10
Maize meal	10	60	0	25	10	50	10	50	10	40
Maize gluten meal	0	10	0	10	0	10	0	10	0	5
Oat meal	0	10	0	20	0	20	0	20	0	0
Sorghum meal	0	40	0	25	0	50	0	50	0	50
Wheat meal	0	50	0	40	0	50	0	50	0	50
Bean meal	0	10	0	15	0	15	0	15	0	10
Cottonseed meal	0	10	0	10	0	10	0	10	0	10
Groundnut meal	0	10	0	10	0	10	0	10	0	10
Rapeseed meal	0	5	0	10	0	0	0	0	0	5
Soya-bean meal	0	40	0	25	0	40	0	40	0	40
Feather offal and blood meal	0	5	0	5	0	10	0	10	0	10
Fishmeal	0	10	0	0	0	10	2	10	0	10
Herring meal	0	10	0	0	0	10	0	10	0	10
Meat and bone meal	0	10	0	10	0	10	2	10	0	10
Dried grass	0	5	0	10	2.5	5	2.5	5	0	0
Tallow	0	5	0	0	0	3	0	3	0	5
Brewers' yeast	0	5	0	5	0	5	2	5	0	5
Molasses	0	2.5	0	5	0	5	0	5	0	2.5

(Adapted from Ministry of Agriculture Bulletin 174.)

'running costs', as it were, of any poultry operation. In the previous sections we have explained the fundamentals of diets for poultry. Now it may be helpful to look at the ways in which a diet can be put together.

Table 4.1 shows typical nutrient requirements of some common poultry diets including the vital factors of protein, energy, the most essential amino acids, vitamins and minerals. This is the foundation on which the diet is built. Table 4.2 gives the composition and nutritive value of some of the commonest foodstuffs used in poultry feeding, whilst Table 4.3 gives the range of normal inclusion rates of feeding stuffs in practical poultry diets.

As well as satisfying the importance of actual and absolute value of the contents there must be a correct ratio between the energy and protein which varies depending on the class or age of bird. For example, with laying hens the ratio between the metabolisable energy (kcal/kg) and the crude protein of the diet should be 187:1. If table 4.1 is consulted it will be seen that for a medium hybrid layer the respective figures are 2800:15, which is 186.7:1, which corresponds well. Growers require a ratio of approximately 176:1, chicks 143:1 at first and then increasing to 176:1 at about eight weeks.

In the construction of the ration, consideration must also be given to the content of indigestible organic material. A bird's gut, as we have shown in the earlier section on the digestive processes, must have a fibrous content to maintain its natural functioning. In young chicks this proportion is about 13 per cent rising to 15 per cent in the adult. In the young bird the figure given should be followed quite closely but in older birds a higher proportion is acceptable, even up to 40 per cent. In broilers, for example, where there is a simple demand for one thing – that is quick growth at as low a food conversion as possible – the indigestible organic matter should on no account go much over 13 per cent, but in growing pullets where some restraint to precocious growth and development can be important to the development of a balanced mature bird, the indigestible element of the ration may rise to nearer 20 per cent.

In deciding on the formulation, all data so far given will be required and in addition it will be necessary to have up-to-date costs of all the likely constituents so that a completely economical ration can be made up to the constraints given. In the end a diet has to be formulated that is (a) of minimum or at least favourable cost, (b) satisfies all the standards given, (c) does not go outside the minimum or maximum inclusion rates for any of the ingredients and (d) represents reasonable uniformity of contents from batch to batch. Because the range of ingredients that can be used is so large, an enormous number of calculations are involved in working out all the permutations. The company compounder uses a computer to enable him to do this efficiently and quickly.

There are various ways of setting about a ration formulation. A simple one is as follows. Firstly, decide on the energy level that the ration should provide. Then set the protein level to give the correct ratio between energy and protein, as described, for fully efficient utilisation. Then follow this by checking the correct amino acid levels to give neither deficiencies nor wasteful excesses, and finally set the vitamin and mineral levels.

An alternative approach in computer formulation may be required if the economic level of energy in the diet is not known in advance. Firstly, the nutrients are set at levels appropriate to a unit of energy. Then the weight of mix can be allowed to vary between fixed limits known to result in acceptable energy levels in the diet. For example, the unit of energy might be set at 1000 kcal ME and the weight of mix allowed to vary between 300 and 400 kg. Then the energy range allowed in the 1000 kcal mix would be 3333 to 2500 kcal/kg. The solution would indicate the most economic level of energy to adopt. If it were 2500 kcal (in 400 kg) then the amount of each ingredient in the formula would have to be multiplied by 2.5 to give a mix of 1000 kg.

Commonly used feeding routines are given below.

Broilers

These are always fed *ad lib* from day-old to finishing. A broiler 'starter' ration usually a 'crumb' contains 22–3 per cent of crude protein and may be fed to an age of 3 or 4 weeks. Thereafter a grower 'pellet' is used containing about 20 per cent protein, followed by 'finisher' pellets of 18 per cent protein. Restriction of intake by broilers by giving decreasing periods of light was referred to earlier.

A recent technique which is of great interest to broiler growers is to feed a proportion of *whole wheat*. Feeding up to 30 per cent saves up to 3.5 pence per bird at 49 days without any loss in efficiency. An investigation by Barbro Tegløf at the Agricultural University of Sweden produced the results shown in Table 4.4.

At present in the United Kingdom whole wheat is introduced at 3 per cent from day 13 rising to a maximum of 30–35 per cent on day 45. Such a technique requires a separate bin for the wheat together with an accurate feed proportioner.

Replacement stock for egg production

These are normally fed *ad lib* from hatching to 8 weeks on baby chick

Table 4.4 Addition of whole wheat experiment made by Barbro Tegløf, Agricultural University of Sweden

0–1 week	0% whole wheat	0% whole wheat
1–2 weeks	20% " "	0% " "
3–6 weeks	30% " "	0% " "
Live weight (42 days)	1733 g/bird	1764 g/bird
Feed consumption, total	3247 g/bird	3339 g/bird
Feed compound	2429 g/bird	3339 g/bird
Whole wheat (25%)	818 g/bird	0 g/bird
Feed conversion	1.87 kg/kg	1.89 kg/kg
Carcase yield	69.7%	71.0%

crumbs (18–20 per cent protein) and then on early growers ration (14–16 per cent protein) to 12 to 14 weeks, followed by later growers reduced to 12 per cent to the point of lay (18 weeks). From 2 weeks onwards mash may be used throughout. Grain balancer-rations are also appropriate. It is likely that restricted diets will become increasingly common practice in the future as a means of economising on the escalating feed costs without sacrificing efficiency. Farmers must follow the detailed programme suggested by the breeding companies for their own birds as these are based on careful investigation.

Commercial egg layers

These are normally fed *ad lib* with the exception of any additions of grain. Protein percentages vary from about 15 per cent for birds on pasture in summer, to 18 per cent for high energy intensive rations. Whilst they may be fed with any form of ration, mash is nevertheless favoured for birds housed in cages. Table 4.1 gives the energy requirements of laying birds.

Breeding stock

The parent stock of commercial egg layers are fed in the same way as other intensively kept laying birds. Extra quantities of vitamins and minerals are given to ensure good fertility and hatchability. Rations for high egg-producing strains should contain up to 19 per cent of the best quality protein.

Separate sex feeding
As the potential growth rate of broiler breeders improves it is increasingly important to control the body weight of the parent stock. An increasing number of farms are therefore using separate sex feeding in the production period. This ensures that the male remains in a fit and active condition for longer with advantages in improved fertility, hatchability and lower feed consumption. It also provides the opportunity for feeding separate rations, optimising the diet of males and females to help maintain fertility, egg numbers and egg quality.

The basic principles of separate sex feeding is to exclude the males from the female feed track and provide a separate male feeding system. The normal method of exclusion is a grill placed on the top of the track which controls access by the horizontal distance between the bars, allowing only the females' narrower heads to gain access to the feed (Fig. 4.1a). An alternative method is a solid cover for the track which restricts access vertically (Fig. 4.1b). Male feeders may be a wide pan or tube system suspended about 450 mm from the floor so it is too high for the females. Such a system makes it possible to control the amount of feed given to the male and also provides a different ration with, for example, lower protein and calcium levels.

The overall benefits of separate sex feeding include better uniformity of males and females, improved control of the body weight of males and females, an opportunity to feed different rations to each sex, better control of feed amounts to both sexes, increased fertility and hatchability overall. Also, the maturity of males can be regulated, feet and leg problems can be reduced, and males with the most genetic potential remain in production longer.

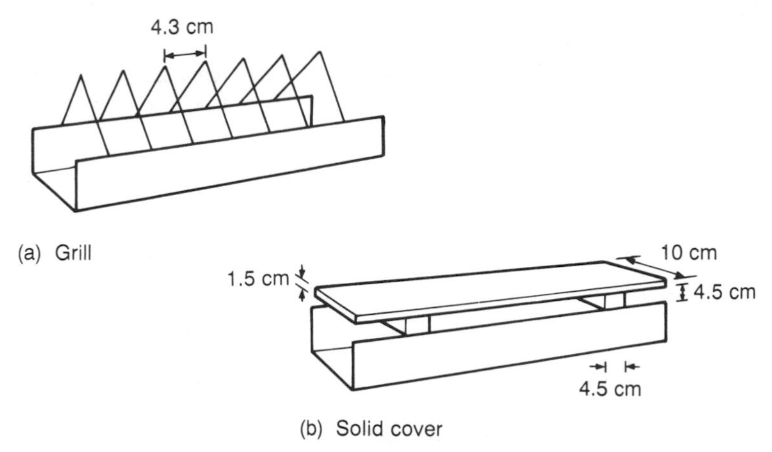

Fig. 4.1 Methods of excluding male birds from the female feed track.

Feeders

Both hand-operated and automatic feeders are available. The space requirements for birds fed using trough feeders are:

1–4 weeks 25 mm per bird
4–10 weeks 50–75 mm per bird
10–20 weeks 75–100 mm per bird.

The requirements for tubular feeders (300–400 mm diameter) are:

1–4 weeks one per 35 birds
4–10 weeks one per 25 birds
10–20 weeks one per 20 birds.

Automatic feeders are mostly trough-types with a continuous chain conveyor taking the food through metal troughs from a storage bin that serves one to four outlets. Another popular scheme is to have a pipe-line delivery which operates by conveying the feed in a closed pipeline to tube or trough feeders.

Though the trough space per bird is important it is just as vital to have a uniform distribution of feeders as it is to have enough space, so that whatever the position of the bird in the house it is not far away from a source of food. This is especially important in the early days of the chick's life, and whilst it is less common nowadays than it used to be, it is still a fact that chicks die because they cannot find water or feed.

The feeders must also be at just the right height so that the food can be readily reached by all birds but not wasted. For the day-old chick a description is given later of the need to place the crumbs on the floor so they cannot avoid them. Later, when they use troughs, this must be gradually raised so that the height is kept about level with the birds' backs.

Drinkers

Poultry drinkers should be automatic, easy to clean, free from excessive splashing, and not liable to flood. The amount of trough space recommended per bird is:

1–12 weeks 12 mm per bird
12 weeks to adult 25 mm per bird.

Table 4.5 Feeding programme and bodyweight guide for typical broiler breeders (Cobb Breeding Company)

Age	Bodyweight			Weekly bodyweight increase	Age (weekly)	Feed/Day (Everyday Feeding)			Ration
Days	g	lb	Actual	%		g/bird	lb/100/birds	Actual	
					0–1	Ad lib	Ad lib		CHICK STARTER
7	115	0.3		187	1–2	to max 35	to max 7.7		18% Protein
14	230	0.5		100	2–3	45	9.9		0.9% Lysine
21	400	0.9		74	3–4	55	12.1		0.42% Methionine
28	580	1.3		45	4–5	64	14.1		ME: 11.80 MJ/kg
35	770	1.7		33	5–6	70	15.4		2820 kcal/kg or
									1280 kcal/lb
									+ anti-coccidial
42	950	2.1		23	6–7	73	16.1		GROWER
49	1100	2.4		16	7–8	78	17.2		15% Protein
56	1280	2.8		16	8–9	78	17.6		0.6% Lysine
63	1430	3.2		12	9–10	78	17.6		0.24% Methionine
70	1545	3.4		8	10–11	78	17.6		ME: 11.23 MJ/kg
77	1655	3.6		7	11–12	78	17.6		2684 kcal/kg or
84	1770	3.9		7	12–13	78	17.6		1220 kcal/lb
91	1880	4.1		6	13–14	78	17.6		+ anti-coccidial
98	1995	4.4		6	14–15	78	17.6		GROWER
105	2095	4.6		5	15–16	78	17.6		No anti-coccidial
112	2195	4.8		5	16–17	78	17.6		**At mating change to**
119	2295	5.1		5	17–18	78	17.6		PRE-BREEDER
126	2405	5.3		4	18–19	78	17.6		16% Protein
133	2520	5.5		4	19–20	90	19.8		0.70% Lysine
									0.3% Methionine
									ME: 11.5 MJ/kg
									2750 kcal/kg or
									1250 kcal/lb
									1.3% Calcium
140	2640	5.7		4	20–21	100	22.0		STANDARD BREEDER
147	2800	6.0		6	21–22	110	243		(As Female)
154	2980	6.4		6	22–23	120	26.5		or
161	3160	6.7		6	23–24	125	27.6		MALE FEED
168	3340	7.1		5	24–25	125	27.6		(if available)
175	3500	7.5		5	25–26	125	27.6		
182	3620	7.8		5	26–27	125	27.6		
189	3720	8.1		4	27–28	125	27.6		
196	3815	8.4		3	28–29	125	27.6		
203	3910	8.7		3	29–30	125	27.6		

The main types of drinkers have ball valves, polystyrene floats, or weight-controlled valves. An alternative arrangement is to have a 75 mm diameter guttering of metal or plastic which runs the length of the house and has water trickling through it continuously. For cages a more usual system is the nipple drinker, which consists of a small rod (or nipple) forming a valve with the metal casing and fixed into a pipe running either along the front or the back of the cage. As the bird pushes the rod upwards a trickle of water flows into its mouth. The advantages are that the cleaning of drinkers is eliminated, the wastage of water is less, and the hygiene is improved.

5 The environmental requirements of poultry

Temperature

Brooding temperatures

A method that is widely used to brood chicks is to arrange a source of warmth in a confined area at about 35°C (95°F) at day-old and then subsequently reduce this by 3°C (5°F) a week. Unfortunately, far too much attention has been paid to the high brooder temperature and not nearly enough to certain other equally important essentials.

In conventional systems there has been a tendency to place too large groups of chicks under a source of heat with a ring of food and water 'outside' this hottest area. This makes it difficult for the chicks in the centre of the group to reach them. As chick crumbs or mini-pellets are placed for easy access on the floor on paper or cardboard it may be the water which is more difficult to find. It is absolutely essential that chick water containers of one sort or another are placed so closely and frequently through thc brooding area that a chick cannot possibly avoid contacting them. Mortality records from the poultry unit at Cambridge havc shown that in a normally managed flock of day-old chicks, on the floor or in cages, the majority of chicks, other than those which are diseased, and which die in the first few days of life may do so from dehydration or starvation. No seasonal effect was found and no evidence of chilling, and good access to food and water is the first essential together with the necessary warmth.

For these reasons, as well as others, brooding systems with larger warmed areas have considerable merit. They enable also a wider distribution of birds in environmentally suitable areas with more space available, factors which are known to improve growth and reduce the likelihood of disease. To ensure a good use of the house, the house temperature is at least as important as the brooder temperature, a range of 25–30°C (75–85°F) being associated with the best all-round performance. Below and above this range, weight gains and food conversion

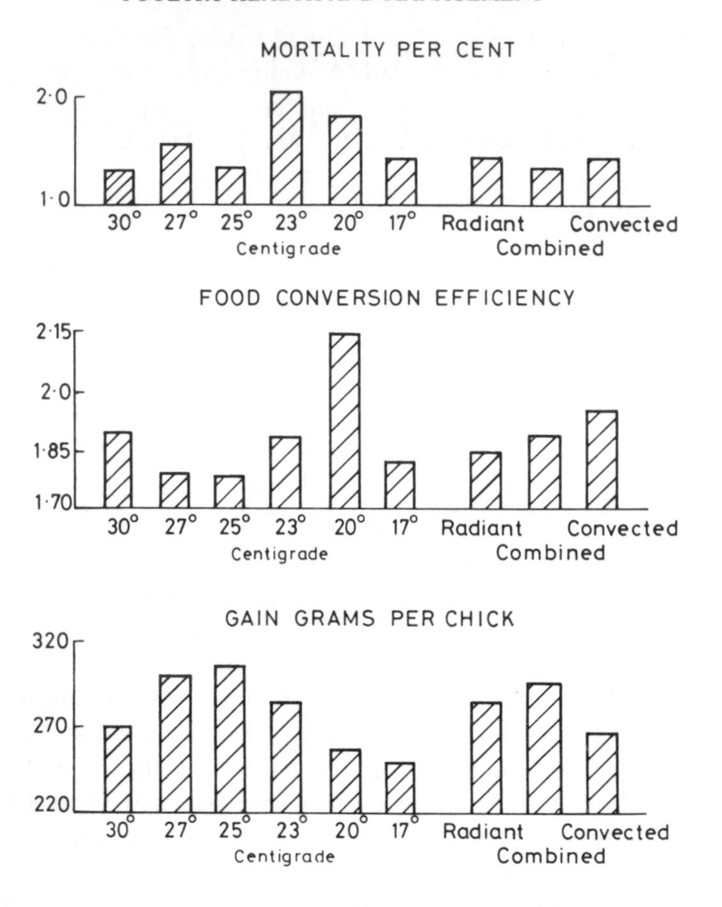

Fig. 5.1 Mortalities, food conversion efficiencies and liveweight gains in chicks kept under different conditions from day-old to three weeks. Note that the best overall results are from radiant heat at moderate ambient temperatures.

efficiencies are reduced. The best performance will probably be obtained if the house temperature is reduced from 30°C (86°F) during the first week to 27°C (81°F) in the second and 24°C (75°F) in the third. The worst results are associated with correct brooder temperatures and low house temperatures, below 20°C (68°F), when the chicks are reluctant to venture away from the brooder heat to feed and drink. On the other hand, too high a house temperature restricts appetite and retards activity and growth. The heat source from the brooder and space heaters should be capable of reaching a maximum of at least 4 kW per 1000 chicks to provide these conditions in well constructed houses. A maximum availability of up to 6 kW per 1000 chicks is preferable.

If the chicks are to be well distributed within the brooding area the temperature must be uniform and draughts at floor level avoided. The conventional but now rather old fashioned arrangement of having solid brooder surrounds about 1 metre high is generally satisfactory in achieving this and still has a place in poor buildings but it has been superseded by warming a larger area and reducing air velocities over the whole house. Overhead, largely radiant sources of heat give the most satisfactory results since their fine thermostat control and adjustable height offer flexible arrangements. They also serve the dual purpose of brooding and space heating. As an alternative, however, blown hot air has its advocates because of its simplicity, low running costs and good space heating qualities. An initial temperature of 31°C (88°F) is recommended in this case which represents something of a compromise between the ideal brooder and house temperature. There should be a reduction of about 0.5°C (1°F) daily until a level of 18–21°C (65–70°F) is reached. All changes should be made steadily and regularly to avoid stress to the birds. Sometimes attempts are made to blow air great distances from end to end of the house which creates considerable draughts and uneven air temperatures. To compensate for this it is necessary to lift the temperature several degrees which can be both costly and unsatisfactory for the productivity and health of the chicks. The correct procedure, however, is to avoid this set of circumstances altogether either by ducting the hot air along the length or width of the house, or so integrating it with the ventilation system that the incoming air is heated as it comes into the house, perhaps through a central intake duct under the ridge taking a mixture of hot and fresh air combined. The very dry air conditions produced by hot-air systems are not entirely favourable to the birds' health and well-being (see chapter 2).

Post-brooding temperatures

From the age of three weeks onwards some further reductions in temperature are justified. In the case of broilers, the house temperature should be in the order of 18–21°C (65–70°F) with a definite tendency to the upper figure if there is any danger of the temperature dropping below 18°C (65°F) owing to external conditions. Under ideal conditions the best growth takes place between 18–21°C (65–70°F) and certain broiler growers are already effecting this by taking great care that reductions to this range are done very gradually, as already mentioned, so that the change is apparently imperceptible to the birds and at the same time ventilation is maintained without a draught. However, in many cases the heat source does not allow the poultryman to adjust the temperature so

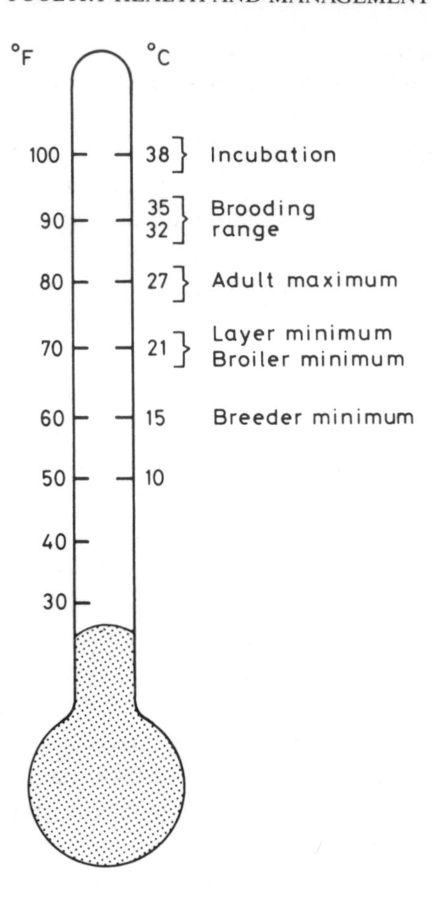

Fig. 5.2 Optimum temperatures for chicken of different ages.

finely or gradually whilst sometimes the ventilation arrangements are inadequate with the higher air velocities demanding a higher temperature to compensate for the cooling effect. Nevertheless, if the ventilation conditions can be maintained satisfactorily, a reduction of some 6°C (11°F) between three to nine weeks, giving an eventual temperature of 13–16°C (55–60°F), is desirable for optimum growth. These temperatures tend to be somewhat lower than those in common practice but their maintenance has important repercussions in reducing heating and ventilation costs and cutting down the effect and incidence of respiratory disease. Obviously such low temperatures cannot be maintained in summer but this is compensated for to some extent by increasing the ventilation and air velocity, if the fans have the capacity to achieve it.

Temperature for layers

For intensively kept birds the optimal temperature is high, that is, about 21°C (70°F). At temperatures below this there is a depression of about ½ egg per hen housed per year per 0.5°C (1°F). Feed intake will be reduced by about 7 grams per bird per day for a rise from 15 to 21°C (60 to 70°F). On the debit side, there is some depression in egg weight, estimated to be about 1 g per egg per 3°C (6°F) rise over 15°C (60°F) but this is far outweighed by the benefits and at the time of writing it is estimated that if a house is kept at 21°C (70°F) rather than 15°C (60°F), then the potential saving in terms of profit margin could be as much as a 20–30 per cent increase in profit.

It is as well also to emphasise that the hen is reasonably adaptable and tolerant to environmental changes and there is a wide range within which it can produce economically even if not at an optimal rate. This range is about 5–24°C (40–75°F). However, this does not mean that the temperature can swing rapidly between these two extremes since rapid changes of any sort are undesirable. Rather it represents the seasonal extremes one should aim for at the outside in designing the housing in the case of less intensive systems than the battery, such as deep litter and the straw-yard. Provided the changes take place gradually birds can acclimatise themselves to them. For short term variation, as between day and night, a maximum of 6°C (11°F) is a good aim.

If the temperature rises above 24°C (75°F) for lengthy periods the total number of eggs laid and their weight and quality will certainly suffer. Appetite will also fall. Below 5°C (40°F) the chief effect will be a sharply rising appetite, though egg weight and quality can benefit slightly. It may be possible to compensate for the falling appetite at high temperatures by increasing the essential nutrients in the ration and so produce as many eggs of almost as good quality on a reduced and hence more economical quantity of food. At present, however, compounders' rations are geared to the 5–24°C (40–75°F) range.

In order to obtain the required temperature conditions in a temperate climate such as the United Kingdom, it is usual to rely upon an excellent standard of thermal insulation and controlled ventilation (see chapter 6). During cold spells it *may* be worthwhile to have supplementary heating to ensure a minimum of 5°C (40°F) in deep-litter houses but it is unlikely to be necessary with commercial layers in multi-bird cages since their body heat will enable a temperature difference of up to about 20°C (30°F) to be maintained – or sometimes even more. On the other hand, with deep-litter housing and especially with breeding stock where the density of stock is very low, the combined heat from the stock and the litter as well can maintain a temperature inside the house often only about 6°C (11°F) above the outside. Experiments have been proceeding for some years to see if there are economic advantages in placing heaters in breeding houses. From records I have been able to study in a number of breeding and commercial egg-laying deep-litter houses, I would say that the evidence favours the use of heaters in the former but not in the latter. However, as fuel costs are constantly rising the wisest course for a poultry owner to take where heat loss problems are occurring, is to take appropriate measures to improve the thermal insulation and tighten up on ventilation control.

The humidity of the air

There is no reason to have a dogmatically set range for the humidity of the air though in practical terms the aim in the winter months will be to keep the relative humidity below about 80 per cent of saturation and preferably nearer to a maximum of 75 per cent. If the advice given in the section on thermal insulation and ventilation rates is followed, this should be possible. In the normal way there need be no problems with chicken if the relative humidity is low, but this is not true if the birds are suffering from respiratory disease (see chapter 11), and thus the evidence is that humidities below 50 per cent to some extent, and especially if they go as low as 30 per cent, may aggravate infection and help contagion. Infectious particles stay suspended and viable for longer periods in dry dusty air. It is mainly for this reason that higher ventilation rates are advised for summer nowadays because while it is not usually practicable to consider humidifying the air to deal with this problem, the dilution of infection and prevention of high dust content of the atmosphere can be dealt with indirectly by high air change rates.

Water cooling

Problems sometimes occur in temperate climates, and often in hot climates, of buildings becoming overheated either momentarily or for quite long periods. This may sometimes arise from poor thermal insulation of the building or there may be insufficient fans or they may be operating inefficiently. If these obvious faults cannot be corrected, or in cases where after this it is still too warm, there are several ways of using water to cool the building.

A simple device is to spray water over the roof and walls, thereby cooling by evaporation; in some cases a perforated waterpipe is placed along the roof ridge to discharge water uniformly along the length of the roof. A more usual way is to pass the incoming air through wet pads; the pads may consist of a wooden frame filled with absorbent wood fibres with water falling through from top to bottom. Surplus water has to be collected and recirculated to be economical and the pads must be kept clean if they are to function successfully.

A still better way is to have spray nozzles which produce a very fine mist. They do, however, need a pump and run-off for surplus water.

A fourth and perhaps the best of all methods consists of a metal disc revolving at high speed which throws off water on to an atomising plate which sets up a very fine mist taken up by the airstream. Good control is achievable by a solenoid valve activated by a humidistat. Full consideration of poultry keeping in hot climates is given in chapter 16.

Lighting requirements

In nature, the development of the reproductive (egg-laying) organ is stimulated by increasing amounts of daylight, as in spring, but is depressed when this is reduced, as in autumn. The modern genetically improved layer, under the stimulus of spring-like conditions, will lay before sufficient bodily development has taken place to fully support egg production and it will not be able to lay either the number of eggs or the larger sizes of which it would later be capable. An autumn-like pattern, or even a constant day length, will allow the body to develop properly before the bird starts laying. Thereafter, to stimulate maximum production, the procedure is to give a weekly increase of light duration of about 20 minutes up to a maximum of 16 to 18 hours. Artificial lighting is of course essential if this is to be achieved at all seasons, though by rearing chicks in the autumn, the natural advantage of seasonal changes can be made use of.

There are a number of techniques used in order to get the most favourable response and most are rather similar but each breeder tends to suggest something different for his own stock, based on sound practical experience. Two programmes are given below for a well-known commercial hybrid. It is noteworthy that the maximum amount of light in one day is given at 18 hours. However, many poultrymen prefer to go no further than 16 hours of light in a day, keeping in reserve the last two hours so they can give an extra boost if egg production for some reason shows signs of tailing-off.

A suggesting lighting programme for commercial hybrid layers is:

0–1 week 18 hours light, 6 hours darkness.
2–18 weeks 6 hours light, 18 hours darkness.
19–22 weeks Increase light by 45 minutes per week to give a good stimulus at the first period of laying.
23–49 weeks Increase light by 20 minutes per week.
49 weeks onwards The lighting is kept steady at 18 hours light per day.

Those retailing eggs and seeking especially large eggs can use the following variant:

0–1 week 23 hours light.
2–18 weeks Decrease by 45 minutes per week.
19–22 weeks Increase by 45 minutes per week.
23–48 weeks Increase by 20 minutes per week.
49 weeks onwards Retain at 18 hours of light.

A new lighting technique that has recently been developed is the use of ahemeral lighting cycles. These are daily light cycles greater or less than 24 hours. These are not capable of altering egg output but they can improve egg weight and shell strength so that there can be economic advantage to the farmer if prices for larger eggs are suitably favourable. The number of second quality eggs can also be reduced. A 28 hour light cycle which uses bright and then dimmed lights has real advantages. The dim lights enable egg collections and stock inspections at any time even during the birds' 'night' period; the bright lights have 30 times more intensity than the dim and this simulates 'day' (bright) and 'night' (dim). A suitable ahemeral lighting programme as devised by the Poultry Department of the North of Scotland College of Agriculture is shown in Figure 5.3.

For broilers the usual pattern throughout most of the industry is to have 23 hours lighting and 1 hour darkness in each 24 hours, the latter being necessary to train the birds to darkness. If this is not done and the light is suddenly withdrawn for some reason, a pile-up is a likely consequence, the birds tending to crowd into corners and suffocate.

There is, however, an increasing interest now in growing broilers on the intermittent lighting patterns, which were once popular, as they appear capable of some improvement in growth rate but more particularly in the food conversion ratio. They are capable of improving digestion with

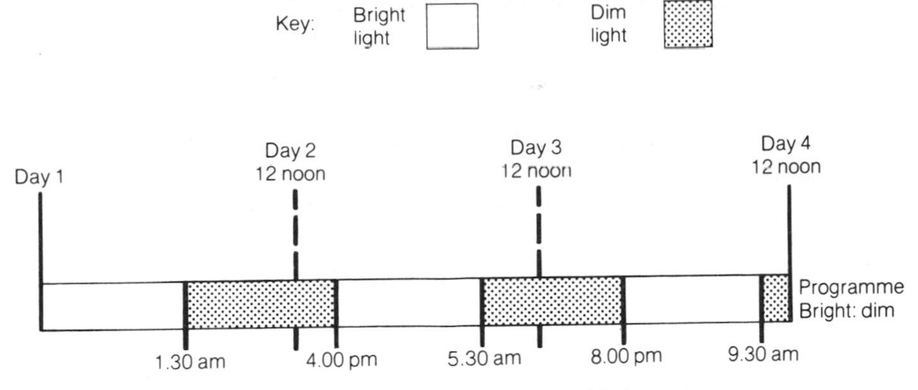

Fig. 5.3 Lighting programme for the 28 hour ahemeral light cycles.

suitable rest periods, decreasing activity and reducing 'boredom eating'. There will also be a reduction in electricity usage which will result in at least a marginal saving.

A suitable lighting pattern would be as follows:

0–3 weeks Continuous lighting (with 1 hour off in 24).
3–5 weeks cycle 3 hours on and 1 hour off.
5–7 weeks cycle 2 hours on and 2 hours off.
7 weeks onwards cycle 1 hour on and 3 hours off.

If intermittent lighting programmes are used it is vital that the highest amount of feeders and drinker space availability must be provided as the pressure will be much greater on both than with continuous programmes.

Continuous lighting for broilers can be with a light intensity as low as 0.2 lux, which is about as low as the 'off' phase in a controlled environment house in the day time.

The lighting procedure

For proper artificial control of lighting, exclusion of all natural light must be complete. Fully efficient baffles under fan shafts or in air inlet hoods and around the edges of the ventilators are essential. Various techniques are shown in the chapter on ventilation (chapter 7) and it is pertinent to stress that methods used to baffle light entry also serve as draught and wind baffles and vice versa. The insides of the ventilators should be painted black. The maximum safe level of intensity of stray light entering houses which are supposed to be blacked out is 0.4 lux and it is advisable if there are any doubts on this score to check the intensity on a light meter.

For stimulation of layers the light intensity should be 10–16 lux and there is no advantage from raising the level above this.

If light intensity is uneven in the house, with bright and dark areas, the birds will favour and concentrate in certain areas. This will tend to cause the development of vices and diseases, particularly respiratory ones. In every poultry house lighting circuit a dimming device is necessary, so that light intensity can be lowered easily should there be an outbreak of cannibalism. If birds are to receive sufficient and even light intensity, the disposition of the lights must be uniform. This uniformity is usually obtained by suspending ordinary tungsten bulbs at about 3 metre (10 ft) centres along and across the house. A reflector over the bulb will assist in maintaining an even intensity and will help to keep the bulbs free of settled dust.

Suitable systems of lighting are now marketed which utilise fluorescent lighting tubes with dimming arrangements and whilst the capital cost of these is somewhat greater than tungsten bulbs, their increased efficiency enables a substantial saving in running costs which can soon recoup the higher initial charge.

6 The poultry house

Construction and insulation

In most poultry houses where environmental control is essential, the first need is for a high standard of thermal insulation of the surfaces which can be absolutely ensured throughout its period of use. First and foremost, a high standard of insulation enables heat conservation to be practised satisfactorily keeping the ambient temperature right for the birds and eliminating the use of food as a fuel – or reducing artificial heat inputs. It also permits a levelling out of any wide diurnal and seasonal temperature variations which may occur and, in association with good ventilation, allows a relative humidity below 80 per cent to be maintained. If the relative humidity does consistently rise above 80 per cent, condensation will occur on the inside surfaces and litter will become caked and wet and serious deterioration of the building may take place. Under such conditions it will be difficult to maintain good health as there will be an increased danger of respiratory disease and, where built-up litter is used, of parasitic, bacterial and especially fungal infections.

The proper way of applying thermal insulation to the building is very important but frequently neglected. Many different materials can be used but it is most essential that these are kept dry. Apart from the fact that many wet materials deteriorate, they also lose their insulating qualities and raise the humidity of the atmosphere. And whilst the importance of weatherproofing is commonly realised, it is not by any means so fully appreciated that it is just as vital to seal the inside from moisture penetration from *within* the building by placing a vapour seal on the warm side of the surface. Suitable vapour seals are polythene, impregnated kraft paper, metal foils, and a number of different special paints and liquid sealing compounds. It is also essential to place a damp-proof course at the base of the walls of poultry houses as with other buildings.

Materials that are used for the inside and outside surfaces should be hard-wearing and maintenance-free. For example, a suitable roof construction satisfying all the essential requirements could be made up from the inside to the outside as follows:

1. Flat, fully-compressed fibre-cement sheets.
2. A vapour seal of well-lapped polythene sheets.
3. 100–150 mm of glass or mineral wool.
4. An air space so there is at least sufficient room to ensure there is no compression of the wool.
5. An outer cladding of corrugated compressed fibre-cement or metal.

It is now becoming popular to use an inner lining of polyurethane or polystyrene boards – which incorporate suitable vapour seals – either because the substance itself is a barrier or because it incorporates a sealing material on the warm side. Increasing application is being made of forms of polyurethane insulations sprayed on as linings or into cavities to set into a material which is about as hard and as serviceable as the made-up boards.

Similar methods are often followed for the walls, though in this case the outer cladding is more usually of timber for appearance sake. Whilst timber boards can be used satisfactorily, suitable alternatives are exterior grades of plywood or oil-tempered hardboard. Certain constructions are able to bond the inner cladding, vapour seal and insulation together as, for example, polyurethane faced with aluminium foil or plastic. The final step of bonding the whole wall or roof construction into one place is actually achieved with the use of polyurethane faced both sides with corrugated steel, aluminium or hardboard usually being used for the exterior surface. There seems little doubt that integrated prefabricated processes like this will become increasingly used in poultry house construction and assist the farmer a great deal by giving a clean hygienic surface.

Recently, poultry housing has made an appearance with walls clad entirely in plastic sheets inside and out with vapour-sealed insulation between the linings. The advantages of using plastic claddings include the absence of maintenance, improved insulation and a generally smarter appearance. However, the initial cost is increased.

The importance of installing insulation properly cannot be overstressed and indeed many potentially excellent houses are ruined by careless finishing. For example, loose fill insulation must be fixed so that it cannot slip to leave large areas uninsulated. Also, joints between boards must be sealed so that moisture cannot penetrate between them.

The floor is also of critical importance yet is frequently badly neglected.

Even with a layer of litter, an earth floor can be an obvious place for the build-up of bacteria and parasites. The dangers are greatest to young birds. There is no good alternative to a simple but adequate concrete floor 75–100 mm thick with a damp-proof course between the concrete and the rubble to prevent moisture penetration by capillary attraction from the earth. Careful consideration should be given to the necessary slope of the floor to give good drainage, especially to assist when the building is being cleaned out. In deep litter systems it is not usually practicable to place drains within the house itself because of blockages but in cage rearing or laying houses internal drains are of great benefit.

Thermal insulating standards

Apart from the strictly practical aspects of insulation it is also helpful to know how it is possible to assess the relative insulation values of different materials or forms of construction. This will help considerably in choosing the best materials and putting in the right thicknesses.

Every material has a thermal conductivity, known as the '*K*' value. This figure is the measure of a material's ability to conduct heat and is the amount of heat in watts flowing through a square metre of the material when a temperature difference of 1°C is maintained between opposite surfaces of a metre thickness. It is a useful way to grade materials according to their insulating qualities and it goes some way to answer the question as to which are the best insulators. The lower the figure the better the insulating qualities. A table of *K* values is given in Fig. 6.1 and it can be seen from this that one would need nearly twice the thickness of mineral wool to give the same insulating value as a given thickness of polyurethane. Nevertheless it may be better and more economical to use the poorer insulator in much greater thickness, depending on the costs of the finished construction.

K values have limited use because surfaces of poultry houses are generally composite structures. What is wanted, therefore, is the rate of heat loss (or heat gain during very hot weather) through the whole structure rather than just the individual materials. This gives much more information than the *K* value and is known as the '*U*' value. It is the amount of heat in watts that is transmitted through one square metre of the construction from the air inside the house to the air outside when there is a 1°C difference in temperature between inside and outside. It is possible to build up the *U* value of a complete wall or roof structure if one has the *K* values of the individual materials (plus one or two other figures) but it is unnecessary to do this as *U* values for complete structures are available from the manufacturers or from technical publications.

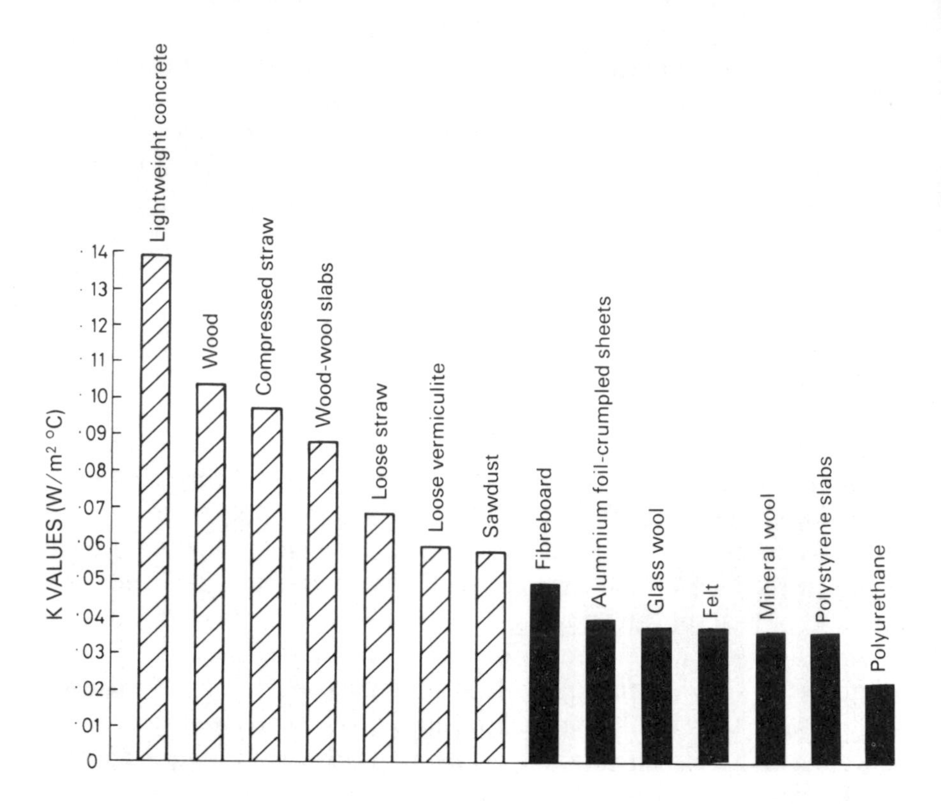

Fig. 6.1 '*K*' values of a series of materials used for thermal insulation. The sizes of the blocks are proportional to heat losses through them so that materials at the bottom (shown as solid blocks) are the best insulators. Note, however, a greater thickness of a poorer insulator can readily compensate (see text).

As with *K* values, the lower the *U* value the better the insulating qualities. A table of representative values is given in Fig. 6.2. The striking reduction in heat loss obtained by insulation is clearly seen, the black blocks being poor insulators and the cross-hatched ones are either reasonable or good. These can be used to set a standard. It is wise in most poultry houses to aim for *U* values in the roof of 0.40 W/m^2 °C or less and indeed it is not an extravagance to look for this in the walls as well. The cost of the insulating material itself is modest and exceptionally good value in view of the very great saving it will give rise to in food costs alone. Doubling the standard of insulation in a poultry house need not add more than 5 per cent to the overall cost of the building.

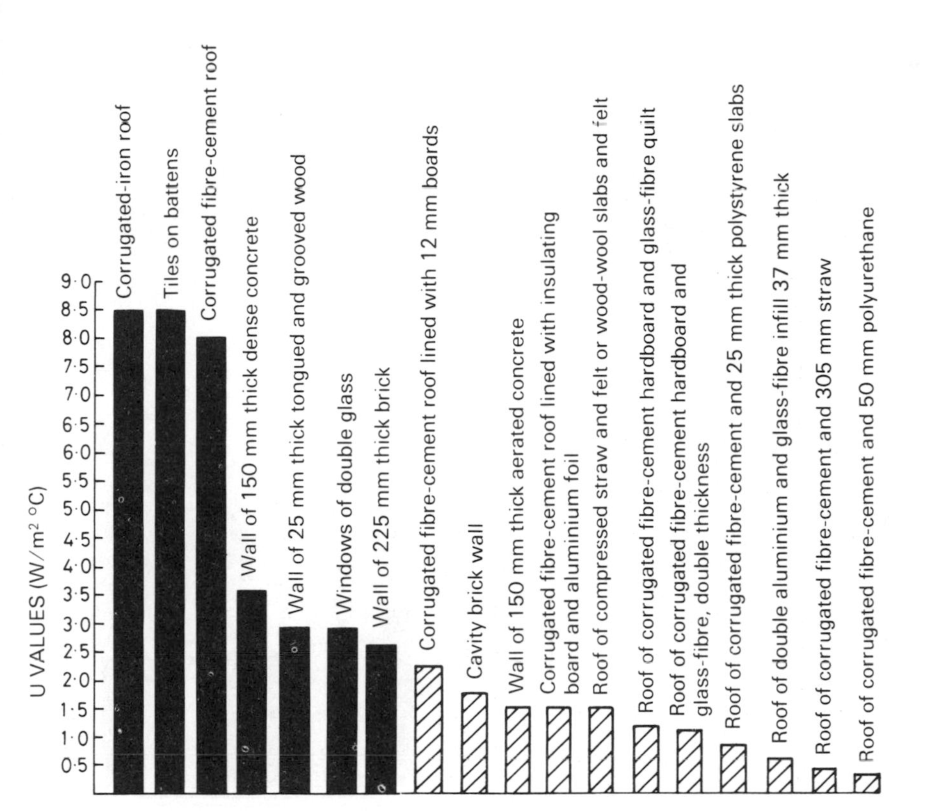

Fig. 6.2 '*U*' values of bad (solid blocks) and satisfactory or good (cross-hatched blocks) roof and wall constructions.

7 Ventilation

Ventilation is concerned with almost all the individual items of the climatic environment, or strictly, the micro-climate. It must eliminate the by-products of respiration and excretion of the bird, and evaporations from the droppings and its litter. It is also concerned with the control of temperature in the house and humidity as we have emphasised earlier. Also, it has to change the air within the building so that the speed of the air's movement is uniform throughout the house. Considerable variation is required from winter to summer. In winter a speed of 0.15–0.25 metres per second (30 to 50 ft per minute) is correct, whereas in summer a movement of at least five or six times that figure is called for. This point will be further emphasised in a practical manner later in the chapter.

Most of the modern developments in intensive poultry management have been concentrated on the so-called controlled-environment house, though it is frequently emphasised that such control as this title implies is but rarely achieved! However, a controlled-environment house need not regulate all the elements in the micro-climate. What is needed is control over the ambient temperature, air movement, ventilation and light. The standard of construction, thermal insulation and ventilation must be high but the cost per bird is brought down to an economic level by stocking at a relatively high density. High densities of stock put a very great burden on the ventilation and the design skill required to achieve all these aims must be very high. The ventilation rates are given in chapter 2 and reference should be made to these for the required figures.

Practical ventilation

There has always been considerable debate as to which system of ventilation should be followed in poultry houses – whether the air should be extracted at the roof and brought in at the sides, or forced in under pressure by the fans to find its way out at the sides or be pulled across the house, from end to end, or from roof to floor. I would always prefer that emphasis should be given to the fact that it is not so much the basic system to be used that is important as whether it is a proper design of its type and is well controlled and operated. Ventilation systems should be kept simple and easy for the stockman to control. Arrangements must always be available for a system that will function should the fans or electricity fail, either by designing it so that natural convection can take over temporarily, or by the provision of an alternative emergency power supply.

It may be helpful first of all to detail the essentials of all ventilation systems. The incoming air must reach the birds at a low velocity in the cold weather; it should also be at the ambient air temperature when it moves over the birds at which time a uniform air velocity should be maintained. This last point is important to ensure an even distribution of birds on the floor, particularly if they are already stocked at a maximum density.

With the most common ridge extraction systems an inlet velocity of 60

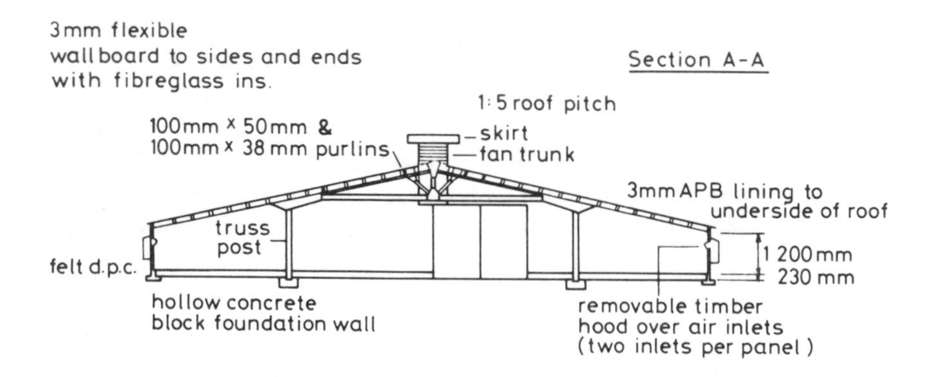

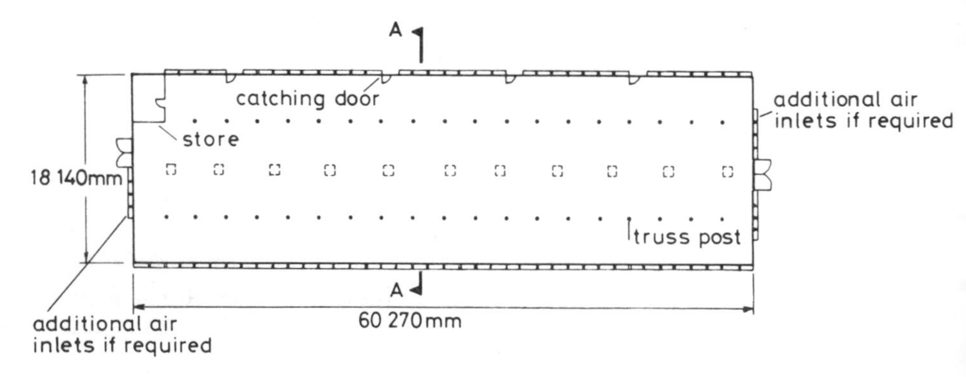

Fig. 7.1 Ridge extraction system with side inlets.

metres per minute (200 ft/minute) is achieved by allowing 0.5m^2 (5 ft^2) of inlet area for each 1700m^3/hour (1000 ft^3/min) extracted. The air inlet velocity is possibly even more important than the air inlet direction but where the air comes in at the sides of the house the direction of entry should be controllable; and when the ambient air outside is cooler than that inside, the air should always be directed upwards away from the floor. If the air enters under the direct pressure of the fans, baffle arrangements should be incorporated to deflect it from causing direct draughts on the stock. If these requirements are carefully attended to, inlet areas as low as 0.18 m^2 per 1700 m^3/hour (2 ft^2 per 1000 ft^3/min) are perfectly satisfactory.

The air should enter or leave the house evenly around the walls or along its length; there must be no dead spots created by large gaps between the ventilators.

The ventilation system should be able to cope in a semi-automatic way with the extremely wide range which occurs between maximum and minimum ventilation requirements, a ratio of up to approximately 100 to 1. For example, in a broiler house where a finished bird requires 7 m^3 of fresh air/hour in summer, a day-old chick in winter requires no more than 0.08 m^3/hour.

In order to obtain the fine control obviously required, it is usual to have a number of fans and speed controls. Thus the fans in use may be progressively increased in number and speed from the minimum demands needed for the young chick. There are a number of acceptable arrangements to choose from; the important features to look for are that usually all fans are speed controlled and that there is a system for the independent operation of individual, or groups, of fans. In order to economise on costs many poultry farmers try to dispense with these refinements, but this is a most unwise economy. With multi-fan systems it is quite possible, however, with certain designs, to avoid using speed regulators by the use of controls that switch the fans on and off in series. The great advantage of systems like these is that they are economical in cost, reliable, cheap to run, and efficient.

The passage of the air movement in a poultry house can be studied by using small smoke candles or fumes generated from a chemical such as titanic chloride. The latter gives perhaps the best indication of the air currents because it is without heat, and fumes copiously on contact with the air. Easily used automatic dispensers of indicator fumes are available commercially at very small cost.

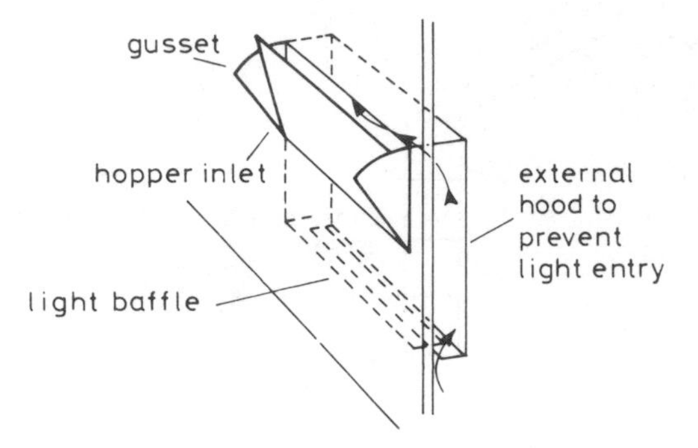

Fig. 7.2 Fresh air inlet inlet incorporating hopper and wind and light traps.

Ventilation systems

The simplest arrangement is to have hopper inlet openings situated along the walls of the building. These are bottom-hinged and open between gussets. The outer hood, with a light and wind baffle, is an essential part. On exposed sites the benefits of the baffle in helping to overcome the effects of strong winds is considerable, and is worth incorporating even when the need for such fine light control is not indicated. In Figs 7.2 and 7.3 forms of baffle are shown. An alternative, more advanced, arrangement is shown in Fig. 7.4. This shows a louvered intake which is set in the inner part of the wall so that there is no obstruction. The louvers give complete directional control which can be arranged from a limited number of points. Automatic regulation is also feasible. Instead of a projecting hood, the air enters the space between inner and outer linings at the base of the wall. This means a tidier appearance, less maintenance and better elimination of light and the unfavourable effects of wind.

A relatively new arrangement is the high speed inlet jet system. The principle of this is to design the inlets so that the incoming air is drawn in at high speed – usually under the ceiling or roof which must be unobstructed. The air stream sets up a number of secondary circular currents of gentle movement which actually ventilate the birds. The advantages are that the small inlet assists in reducing the wind's unfavourable effects on ventilation thereby ensuring that the air movement and distribution are uniform, stable and predictable. Inlet speeds at the intake are usually maintained at least at 5 m/s (1000 ft/min). New control devices have been designed to give automatic regulation of the inlet area and the number of fans operating which must be at full speed or 'off'. This system gives very economical control and is an energy

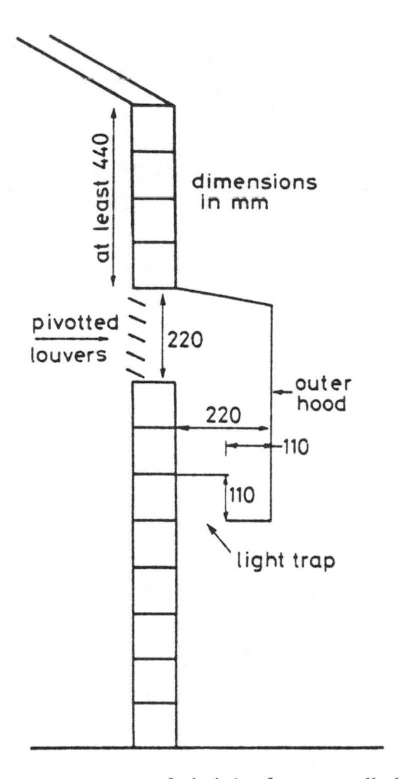

Fig. 7.3 An alternative arrangement of air inlet for controlled environment housing.

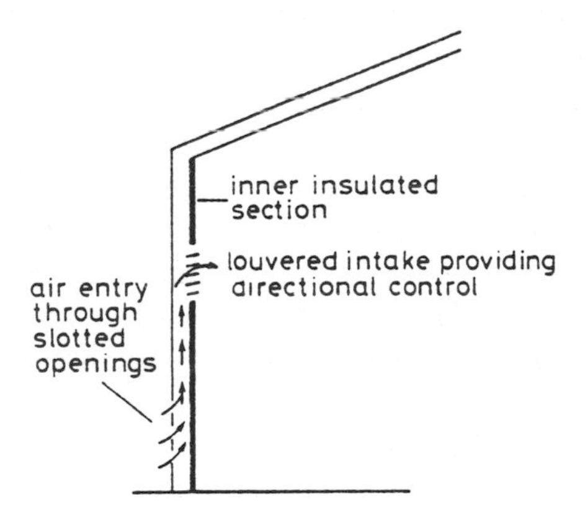

Fig. 7.4 An excellent design of fresh air inlet integrated into the wall structure.

conserver by utilising a minimum number of fans at top speed and by keeping the coolest air under the ceiling surfaces.

With any of these intake arrangements, extracting fans are usually placed in the ridge. Manufacturers' cowls can be used but, as they are

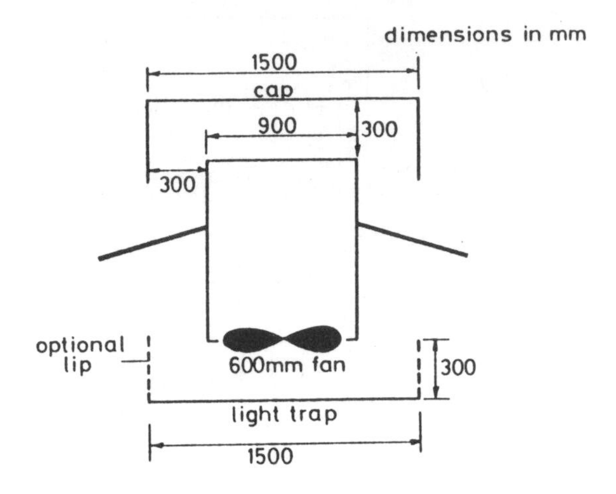

Fig. 7.5 Extracting chimney trunk for mounting on ridge, with fans and light trap.

rather expensive, the more usual procedure is to place the fan in a square timber trunk. The trunk should be larger in area than the fan, but a diaphragm fitting round the blades is vital. A light trap is needed underneath and the cap over the top also has a lip to assist in keeping light out. Typical dimensions are shown in Fig. 7.5. A chimney trunk of 750 mm × 750 mm (30 in × 30 in) is used for a 600 mm (24 in) diameter fan. The fan is placed at the base of the trunk in its diaphragm plate – this is, in fact, the most efficient placing and it is also easiest for maintenance. In most poultry houses 600 mm fans are used as they are the most economical, though there are special uses for 750 mm diameter fans and for 450 mm and 350 mm diameter units. If there is a power failure, the trunks serve as straightforward static extractors and save the house from becoming disastrously hot and under-ventilated. In any sizeable unit a stand-by generator is best to cope with power failure and with most other systems of ventilation it is an essential part of the enterprise.

The conventional system of ventilation described here remains suitable for buildings up to about 16 m (50 ft) span whether the fans are placed for ridge extraction with the inlets on each side of the house or for cross ventilation with the inlets on one side only and the extracting fans on the opposite side. The system is also suitable for birds in tiered cages when up to three rows are used. However, with four or more rows, the outer cages tend to produce a barrier preventing air from reaching the centre and make additional inlets necessary. One way of providing them is either to place ducts under the floor to open under the central cages (Fig. 7.6) or to raise the whole floor off the ground so the air can enter under the

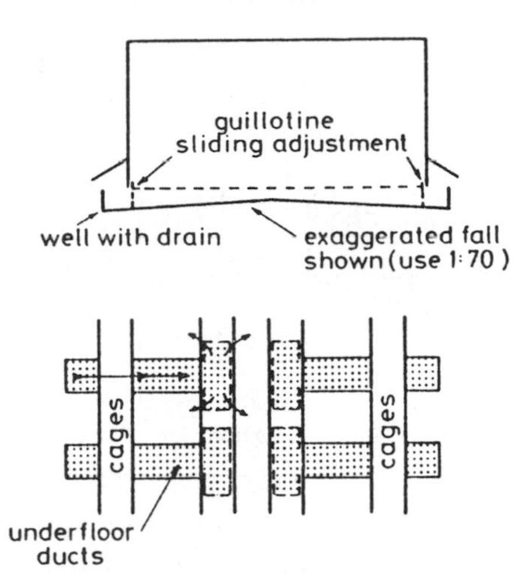

Fig. 7.6 Under-floor ducting for entry of fresh air under battery cages.

central rows. Alternatively, ceiling inlets by strip-gaps or areas of glass-fibre which bring air down between the central rows can be installed (Fig. 7.7). Yet a third arrangement has been used very successfully in cage laying houses of up to approximately 30 m (100 ft) in length with the air being blown from end to end. It is pertinent to emphasise that in all cage laying houses, a satisfactory air flow is more likely to result if the passages between the cages are adequate. A 1.3 m (4 ft) passageway is ideal in this respect.

In certain cases, such as wide-span houses and deep-pit houses with stair-step (Californian) or flat-deck cages (see Chapter 9), the air is

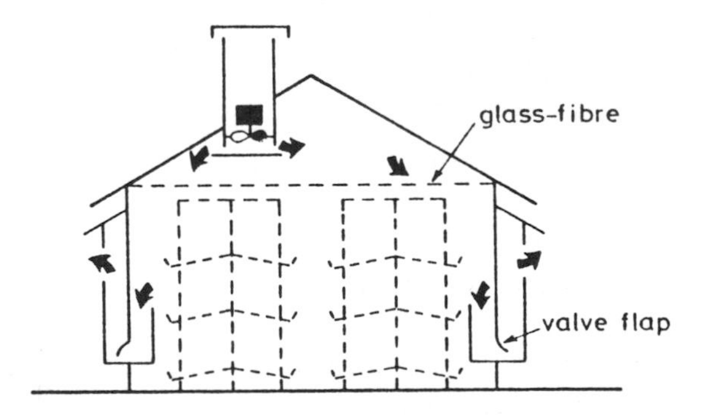

Fig. 7.7 Pressurised reverse-flow ventilation, air entering either through ceiling strip intakes or glass-fibre.

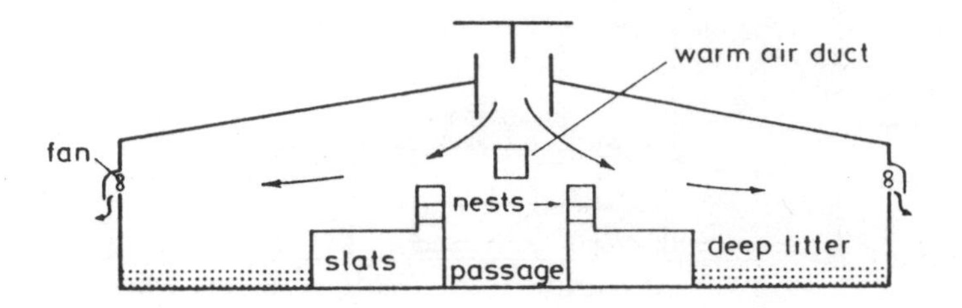

Fig. 7.8 Reversed air-flow with ridge intake suitable for wide-span and Californian cage systems.

brought in through the ridge and extracting fans may be placed in the wall. The design of the inlet at the ridge is of great importance to prevent cold down draughts on the birds. One way of avoiding the draughts is to place inlet chimneys in the roof (Figs. 7.8 and 7.9) with the air passing into them at not more than 243 m per minute (800 ft/min) (inlet area about $0.054m^2$ ($1.5\ ft^2$) per 1700 m^3/hour (1000 ft^3/min) extracted). A cheaper arrangement is to use an open ridge with a baffle board underneath (Figs. 7.10 and 7.11); this may be fixed in position, which is a simple, unsophisticated arrangement, or may be controllable both in area and direction so that different degrees of opening and different arrangements for the deflection of the air can be achieved. Downturned fan boxes, as shown in Fig. 7.10, may be protected from the adverse effect of the wind by the use of a suitable windbreak, as depicted in Fig. 7.11. A more complex technique for air entry is to place louvered openings in each gable end and run a duct of perforated hardboard along under the entire ridge length (Fig. 7.12). This serves as the inlet and diffuses the air. Warm

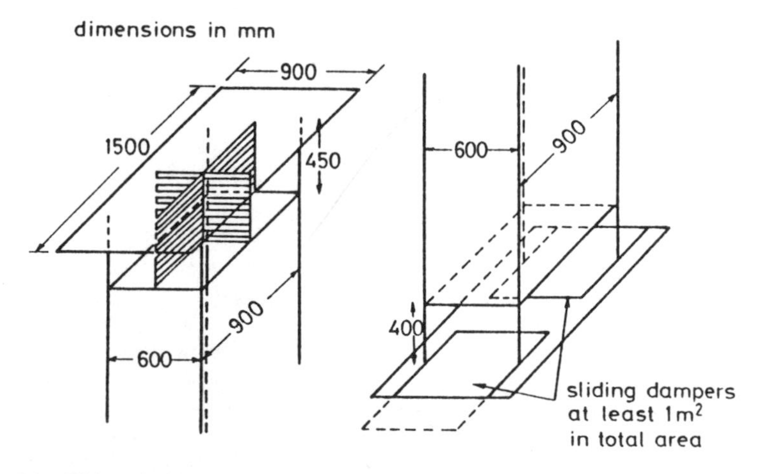

Fig. 7.9 Ridge air-intake.

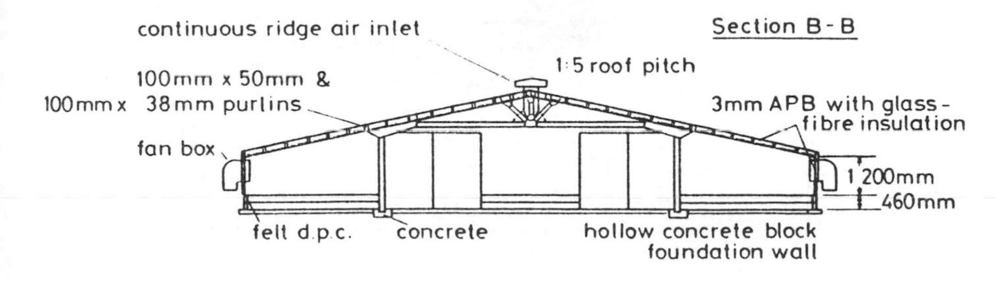

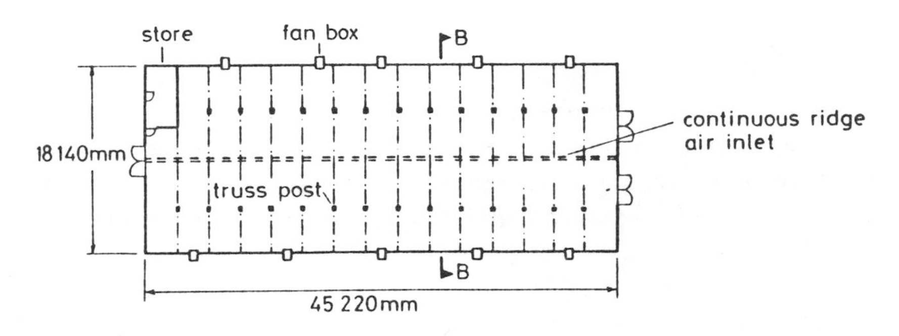

Fig. 7.10 Ridge inlet system with wall extracting fans.

air may be injected into the duct if required; for example in breeding or brooding houses. Great care has to be exercised on the detailed choice of the type of perforated hardboard and the correct size and shape of duct, which can only be determined for each specific form of house. In general, a hardboard duct with as large an orifice size as possible should be used,

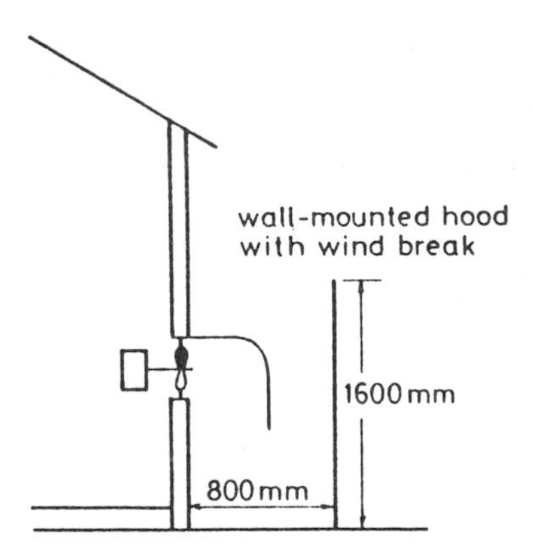

Fig. 7.11 Enlarged detail of wall hood.

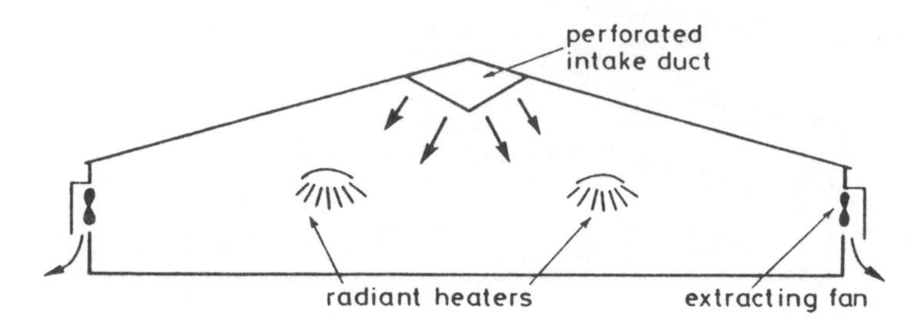

Fig. 7.12 Ventilation by perforated intake duct: louvered gable inlets.

and duct dimensions must be such that air does not pass along it at more than 300 m (100 ft) per minute and preferably less. The main practical problems that arise are from the use of openings so small that they quickly become blocked with dust, or far too large openings that lead to excessive down draught especially in the region of the louvers, or the fans if they are placed in this position.

Air diffusion

A further use of perforated hardboard is to place it over the inlets in the walls to diffuse the incoming air, and also it is a useful way of eliminating the worst effects of wind. It can be placed either across the base of the hood or at the point where the air passes into the house. With this arrangement, as with the use of other diffusing or 'filter' media, great care must be taken to ensure that the holes or fibres do not become clogged with dust or debris, for otherwise the amount of air entering the house may be seriously reduced.

In any system of poultry housing where the droppings accumulate under the birds, such as the deep-pit house, it is absolutely essential to have a reversed flow with the air passing in at the ridge or ceiling, then to the birds and finally over the droppings and out. This keeps the air purer, free of gases, and the droppings dry more quickly. Whilst it is still common practice to achieve this by placing the extraction fans in the pit area, it is now becoming more popular to pressurise the house and place the fans on the input side to create the pressure.

The glass-fibre ceiling

One way of achieving a good reversed system and pressurised ventilation arrangement is to use the whole ceiling of up to 50 mm (2 in) of glass-fibre through which the incoming air is diffused. It has the advantage that the incoming air passes into the house over a very wide area so that the

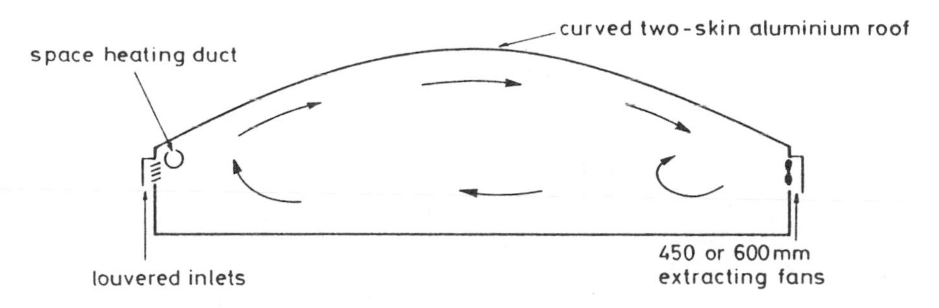

Fig. 7.13 Cross-ventilation used principally in curved roofs of up to 15 m span.

circulation is draught free and uniform. The degree of pressurisation helps to overcome the wind's adverse effects and the reverse flow makes for a more dust-free atmosphere for the operator. The filtration may have useful health effects by some purification. It is important whenever the roof space is used as the pressure chamber (as in this case) that it is well sealed from leakage of air and it is desirable that it is both lined and insulated. Otherwise there is a danger that, during warm and very sunny weather, the roof space will become so hot that the air will be pre-heated before it even goes into the house. Another especial need is to ensure that there is both an alarm system and, or, an automatic standby generator, as otherwise in such a well-sealed house the birds will very quickly suffocate. A suitable arrangement with good baffle outlet is shown in Fig. 7.7.

Figure 7.13 illustrates a very acceptable cross-ventilation arrangement for curved roof buildings in particular of up to about 15 m (50 ft) span. The placing of the fans on the wall has the advantage that they are easier and cheaper to install and service, while an uninterrupted ridge line or roof construction economises on costs. Distribution of air is good and the unfavourable effect of wind is reduced.

A warning must be given. With many of the most sophisticated mechanical ventilation systems, if there is any failure in electrical supply or fan breakdown, there can be little or no air flow. Thus there is a grave risk of the birds suffocating. Measures must be considered to install an alarm to warn of failure, and a standby generator to take over if there is an interruption of mains supply. In the absence of a generator, a fail-safe system should be provided in which a considerable area of shutters normally held together by electromagnets, can open automatically when the electric current fails. Great care is needed in making such provisions and expert advice should be sought for the installation of alarms, generators and fail-safe measures.

Plate 7 For wide-span houses the ventilation arrangement shown is favoured with reverse ventilation pulling in the air at the ridge – in this case an open ridge with control flap underneath – and the extracting fans in the wall.

Plate 8 An even better way of achieving the reverse flow. Fans in the gable end of the roof space push the air over a perforated ceilig to give positive pressure flow and uniform movement of air from ceiling to floor.

Plate 9 A similar building to that in *Plate 7* but of smaller size with outlet arrangements at the bottom of the side walls. The continuous outlet situated along the walls near the floor is protected by weighted polythene flaps on the outside to remove any chance of back-draughts.

8 The disinfection of poultry houses

It has been emphasised in recent years that there is an overriding demand under intensive systems of poultry keeping to have impeccable hygiene. The foundation of achieving this is by careful and proper use of a planned disinfection programme. Disinfection of a building can take place by the process of nature (natural disinfection) and by artificial means (artificial disinfection).

The disinfection of a building implies the elimination from the house of all micro-organisms that are capable of causing disease – thus implying the conversion of a place from a potentially infective state into one that is free from infection. Disinfectants are the means of achieving this and are usually chemical agents. It must be stressed, however, that a total cleansing is an essential preliminary to disinfection. Organic matter has great power to reduce the effectiveness of disinfectants and without prior cleansing the disinfectant's power can be completely invalidated. On the other hand, if a good detergent sanitiser is used, for example HD3 or DSC-1000 (Antec International), it has been found that very nearly all pathogens are destroyed even before the full disinfection programme is applied.

Natural disinfection

Most pathogenic micro-organisms do not survive very long outside the animal body but unfortunately sufficient may always remain to cause renewed infection. Bacteria and viruses can live for several months if protected with organic matter and the spores of bacteria can live almost indefinitely in the soil or protected in cracks and crevices of the building. Even the coccidial oocyst may survive for years in infested quarters.

The factors contributing towards natural destruction of microbes are nevertheless important, and are in themselves worthwhile aids to the artificial processes. Sunlight, heat, cold, desiccation and agitation all contribute. The ultra-violet rays of sunlight are most potent and their powers of destruction are enormous, but unfortunately they have little penetrating power and cannot pass through glass or translucent roofing sheets, nor through cloud or industrial or agricultural haze. The value of sunlight in animal buildings is therefore wholly unreliable. Desiccation from fresh air and wind will also contribute to the destruction of microbes, particularly when the micro-organisms are exposed to these elements by prior cleansing of the building, but in the protection of poultry houses – and bearing in mind the very short period of only a few days between batches of birds – the effect of natural exposure to the elements is as unreliable a 'killer' as is reliance on the sun.

It is also poor economy to leave houses or whole units empty in an attempt to rid them of infective agents: if such a period were to have any meaning it would need to be a very long one indeed, and even then might be a waste of time. So the procedure that is advised (and with justification by results) is to carry out such a thorough cleansing and disinfection that it may be considered safe to place birds back into the building within hours of the programme being completed – once the building is dry and any residuum of harmful disinfectant agents has gone.

For many years heat has been a favoured method of achieving sterilisation and disinfection, particularly moist steam under pressure. In farm use, dry heat is used with a flame-gun and moist heat with the 'steam-jenny' but both tend to be inexact and uncontrolled methods of applying these agents. For example, clostridial spores may survive heating at 300°C (572°F) for 10 minutes. The very transitory heating from a flame-gun would therefore be an uncertain way of attempting to achieve disinfection. Likewise, with moist heat, clostridial spores can survive up to 300 minutes at 100°C (212°F). The uncertainty of steaming is thus apparent, and whilst it is a useful cleansing agent it cannot be considered a reliable disinfectant when used in buildings where the microbes may be protected in cracks and crevices. On equipment it is more useful, but its powers are greatly increased by incorporating a detergent and disinfectant in the steam or hot water. It must again be emphasised that the presence of organic matter will interfere markedly with disinfection by heat.

Chemical disinfectants

Disinfection on poultry farms is generally carried out by using chemical agents. The lethal action of disinfectants is due to its ability to react with the protein and in particular the essential enzymes of micro-organisms. Any agent that will therefore coagulate, precipitate or otherwise denature proteins will act as a disinfectant. Amongst these agents are phenols, alcohols, acids, salts of heavy metals and hypochlorites, as well as heat and certain radiations.

Disinfection is not an instantaneous matter – it takes place gradually; however, many more microbes are killed at the beginning of the process than at the end, though there is an initial lag period before activity commences. Also, the activities of most disinfectants increase as the temperature is increased, though there are a few exceptions.

The effect of organic matter

Almost invariably when disinfection is carried out on the farm organic matter will be present. Its presence will interfere with the action of the disinfectant in the following ways. The organic matter may protect the cell by forming a coating on it thus preventing the ready access of the disinfectant. Also, the disinfectant may react chemically with the organic matter, giving rise to a non-germicidal reaction product. In other cases the disinfectant may form an insoluble compound with the organic matter to remove it from potential activity, or particulate and colloidal matter in suspension may absorb the antibacterial agent so that it is substantially removed from solution. Finally, fatty material may inactivate the disinfectant.

Approval of disinfectants

In most countries official tests lay down rigid procedures for testing and approving disinfectants. For example, in the U.K. the tests cover a wide range of appropriate organisms in the presence of organic matter. The current Approved Disinfectants Order specifies over 300 disinfectants in the various groups. Those groups that are of special importance to the poultry farmer are Group II covering those against fowl pest, and Group IV for general farm use, the test organism being *Salmonella choleraesuis*.

Types of disinfectants

There is a wide range of disinfectants used in poultry husbandry and the following information may help the farmer to make an appropriate choice.

Phenols and related compounds

Derivatives of phenol – which is not itself used – remain as some of the most popular of all for general farm use. They can act well against bacteria and their spores and also against fungi. By the incorporation of organic acids, surfactants expertly blended, their range of activity can also include excellent virucidal activity. For example, Antec Farm Fluid S, produced by Antec International, has additional chemicals, such as propylene glycol, triethanolamine and l-napthyl-N-methyl carbamate increasing activity against insects, mites, worms and coccidia. Antec International also produce Longlife 250S with similar characteristics and Micro-Biologicals produce Macrokil and Microzol with similar actions. These products may be mixed with water or oil and in the latter case the mixture is an ideal treatment for earth floors as the penetration will be good and the residual protection lasts for a few weeks. In practice also it must be emphasised that many houses have badly damaged floors and porous concrete block wall surfaces which will greatly benefit from such treatment. Such products are specially formulated for terminal disinfection when a complete spectrum of activity is required, since it is safe to assume that possibly every type of pathogenic organism is a potential risk and the specific identities of the pathogens are generally unknown.

For use after outbreaks of coccidiosis a disinfectant that generates ammonia must be used as this has specific destructive effects on the coccidial oocysts. One example, Oo-cide (Antec International) achieves this and also has widespread activity against bacteria and other pathogens.

At present the poultry industries of the world are under the greatest threat from highly damaging viral infections. The enormous sites that have been established provide a dangerous focus of infections and disinfectants that have a known activity against all the viral families should be used. For example, Antec Virkon S satisfies these demands. This disinfectant is a balanced and stabilised blend of peroxygen compounds, surfactants, organic acids and an inorganic buffer system. It is readily soluble and safe and also biodegradable. Therefore it can be used in all the following applications: in sanitising the drinking water, in washing out the house, in disinfecting the house, and in 'fogging' the environment when the house is ready for repopulation.

There is also a considerable following for the use of a product such as Virkon S during the lifetime of the bird. This material may be added to the drinking water at a level of 1:500 to limit the spread of viruses and other organisms. Also an aerial spray, using a 1:200 solution of Virkon S over the birds, can be used to reduce the build-up of infection within the house and will assist in controlling the challenge from secondary bacterial infections such as *E.coli*. Approximately one litre of solution is required for each 100 m^3 to fill the house with a fine mist. Specialised equipment is required to do this work effectively, though the choice may be made between knapsack sprayers, pressure washers on very fine settings, or mechanical fogging machines.

Formaldehyde

Some mention must be made of formaldehyde. This is a widely used disinfectant both in the gaseous and aqueous forms, being virucidal, bactericidal and fungicidal. Formalin, which is an aqueous solution of formaldehyde gas containing 40 grams of formaldehyde in 100 millitres of the solution, is widely used at 5 per cent strength as a general disinfectant but it does need to be in contact with the surface for some time to be effective. Its action is greatly affected by temperature and the warmer it is the better – blood heat being most satisfactory. It also acts most efficiently when the surface is wet.

Other than spraying formalin as mentioned above, solid paraformaldehyde granules, e.g. Alphagen prills, are used by placing them on remote controlled electric hot plates at a rate of 1 kg per 360 m^3 of building space. Formaldehyde prills are also available incorporating formaldehyde plus an insecticide, thus increasing the spectrum of activity.

It is uncertain how much longer the use of formalin/formaldehyde will be permitted. In some countries its use has been prohibited as a health hazard to humans. Where it is used, always obey the precautions which are given by the suppliers absolutely.

Control of ectoparasites

It is of great importance to control ectoparasites on the farm; they are a nuisance to both the attendant and the animals. They may be the cause of direct irritation and disease and they may indirectly spread disease – indeed, any infectious agent can be carried by ectoparasitic vectors. These ectoparasites include flies, such as the ordinary house fly, the lesser house fly and the stable fly. Also mites, including red mite and northern fowl mite, ticks, lice, fleas, bugs, beetles and cockroaches. One of the worst offenders is the lesser mealworm or litter beetle, *Alphitobius diaperinus*, which is a common carrier of virtually all poultry diseases.

The most effective control of insects are preparations incorporating organo-phosphorus compounds. For example, Micromite (Micro-Biologicals) or Nuvanol (Ciba–Geigy). Such materials are hazardous to humans and must be used with great care.

Practical essentials

Regrettably it has to be recorded that the way in which the process of disinfection and cleaning is carried out on the poultry farm very often falls far short of an acceptable standard. Millions of pounds must be wasted annually on the misapplication of good disinfectants so that the disease-causing pathogens quite comfortably survive to reinfect the next batch of poultry. Many of the organisms that cause disease are extremely resistant to destruction, especially if protected by even a minute amount of dirt or débris. Salmonella organisms, for example, have no difficulty in surviving for a year or so and even the relatively more fragile viruses, such as those causing virulent Gumboro disease, can survive for months under ordinary conditions. In addition the area immediately around the site, which may contain heavy deposits of infected dust and debris, is often ignored in the cleaning and disinfection process. Indeed it is frequently so neglected in general care that it can be a quagmire of slurry. There are, in addition, the active and living 'carriers' of infective organisms, such as the litter beetle. Few sites seem to be without these – in the warmth of recent summers their presence was greater than ever seen before – and evidence quite clearly shows that they are potential carriers of all pathogenic micro-organisms.

Poultry health and hygiene

To a great extent the poultry industry has thought that most diseases could be dealt with satisfactorily – or could at least be controlled – by the use of medicines and/or vaccines. In many cases this has been reasonably true. Consider some of the most common conditions, such as infectious bronchitis, Marek's disease, Gumboro disease, Newcastle disease, coccidiosis, epidemic tremor, *E. coli* septicaemia – the list goes on – but essentially these are either prevented or treated with vaccine or medicine. Now, however, we are faced with several conditions that cannot be dealt with in such a straightforward way. The new virulent form of the Gumboro virus is not adequately dealt with by vaccination alone and certainly not by medication. The 'nasty' types of Salmonella, such as *enteritidis* and *typhimurium*, will in the former case and may in the latter, be dealt with by slaughtering out the flock. There are also the other zoonotic conditions (diseases communicable to man) where infected poultry, though showing no symptoms themselves, may convey disease to man of quite a serious nature. Three of the most likely are *campylobacter*, *listeria* and *staphylococci* and there are strong public health demands on our poultry farmers to do something to control these conditions. The answer, of course, is to look to good standards of hygiene – in other words the highest standards of cleaning and disinfection. Other vectors of disease, such as vermin and feral birds, are also known to be as dangerous carriers of viral and bacterial infections.

Phases in the procedures

A basic essential is that the cleaning and disinfection programme must be most carefully planned. *All* birds and *all* litter should be off the site before the cleaning begins. The amount of infected material that is stirred up and floats around the site when the birds and litter are removed is so enormous that any cleaned house is re-infected. I would also emphasise that the litter itself is also vastly infected and it should not only be removed from the site and covered while being transported but should be stacked or spread as far away from poultry sites as possible. In the recent outbreak of virulent Gumboro disease we could clearly relate site infection from the spread of infected litter. Thus, the phasing of procedures will involve clear demarcation of the cleaning and disinfection into four totally separate stages:

(1) Birds and litter are moved out.
(2) Wash down with a detergent disinfectant.
(3) Disinfection.
(4) Fumigation.

An overlap between the processes is undesirable especially for the first procedures, which are the more dirty ones and therefore more liable to spread infection. There is very much less danger with the last two processes.

From studies recently completed I have been able to establish that the recontamination of a house that has been disinfected takes place speedily and seriously if the cleaning process is continuing on the site. Indeed it takes place even if there is no cleaning, just from general environmental pollution. Thus, whilst it is important to get the various hygiene procedures in proper sequence, it is also important to put the site into production as soon as the houses have been cleaned, disinfected and dried. There is, in other words, not only no merit in leaving a house empty for a time but the whole concept of assisting de-contamination by having a prolonged fallow period is contra-indicated.

Cleaning

There is no hope of the overall hygiene process being fully effective if there is any 'muck' left in the house. In the real world of the poultry farm there are not many houses or sites where a fully effective job can be done. This is because there will be some basic flaws in the construction, allowing penetration of the structure by pathogens or carriers of pathogens – for example, pervious surfaces, poor protection of joints, wear on floors and inaccessibility of certain parts such as ventilators. There may also be a huge reservoir of potential danger in cavities within the construction. A common area of danger is the cavity between the inner and outer cladding in a prefabricated building. Also, older buildings which are in other respects often satisfactory, cannot meet with the much greater demands of modern hygiene.

We have found repeatedly that when the cleaning with a good detergent disinfectant has been done with great care and the muck has been cleared in its entirety, that there is little sign of any bacteriological and fungal infection remaining. This, therefore, enables the two final processes of disinfection application and fumigation to give the final near-sterilising effect. The problem of course is that in most buildings there are numerous places where it is difficult or even impossible for the cleaning process to do a thorough job. Unfortunately, too, many houses have been built with almost no thought for the hygienic processes. Porous building blocks, poor flooring surfaces, fragile expanded plastic for the roof, dangerously exposed electric wiring and electronic controls all add up to enormous problems. Add to this the almost certain inaccessibility of parts of the ventilation system. Some surfaces of the building will not withstand the application of a high-pressure washer so that the only measure one can take is to blow down all the dust and rely on the disinfectant and fumigation to do the job. As new buildings replace the old, the modern surfaces and construction should enable a much more effective job to be done.

In this general cleaning the surroundings of the site need careful attention. It is clearly best if the immediate area around the houses is concreted so that it can be given the same treatment as the building itself. It is best to look upon it, and also the outside surfaces of the roof and walls, as being a part of the building and requiring the same dedicated treatment.

In those houses with ridge fan or ventilator extraction a great deal of debris accumulates on the roof and this needs to be cleaned off and the site disinfected. If the fans discharge at the side then there are equal, if not greater, deposits on the ground. If it is discharged on to concrete it may be cleaned off but if it is on to the soil in a bed of weeds and slurry then it remains a potentially potent source of re-infection unless it is removed and/or suitably treated. Some years ago there was a vogue for dealing with this problem with a flame-gun but, in careless hands, this turned out to be a rather more effective way to destroy houses than remove pathogens. I prefer the use of good management of the greensward so that the area is attractive rather than being an eyesore.

The disinfection

The house is now ready for the application of the disinfectant. Follow the manufacturer's instructions carefully and make sure that the disinfectant has proven activity against the organisms it is known must be destroyed. These are likely to range from resistant and encapsulated bacteria, such as salmonella or clostridia, to vulnerable viruses such as those which cause Gumboro disease or runting and stunting disease, together with fungi, mycoplasma and others. Special products will be needed additionally for the destruction of coccidial oocysts or for insects such as the litter beetle and vermin. It will not be necessary for these to be used after every crop but, wherever there is a clear indication of infestation.

In several studies I have done in the hard practical world I have found it has been beneficial to use disinfectants at either double the strength or double the quantity. Many surfaces are very absorbent and take the disinfectant away, as it were. In other cases it is impossible to remove all the organic matter effectively and in addition it is well to remember the temperature effect; most disinfectants are far less active in very cold conditions and this may be counteracted by improving the concentration or quantity.

Serious points of neglect that are left alone to the great detriment of the stock are all the equipment, including feed, heating and electrical fittings, feed bins and perhaps above all the entire water system. I have found little difficulty in picking up pathogens at all these points when the usual rather slap-happy disinfection programmes have been followed.

Fumigation

This is the very important procedure to give a final boost to the disinfection programme. At this stage the house is set up ready for the next crop and the fumigation can be carried out by a fine mist with a pressure washer or mechanical fogging equipment. I am afraid the user must be warned that sometimes the most spectacular looking mist fails to get to the parts where it is most needed and it is advisable to check very carefully that a proper job is being done by carrying out some bacteriological tests. For instance, in many cases areas protected by posts and beams fail to get much benefit from this process. What is sought is a coating of disinfectant with as long a residual effect as possible over all the surfaces, to see the birds through the first crucial weeks of their existence in the house. There is no real reason why the house – crop after crop – cannot maintain its healthy status if a thorough programme is applied. The first weeks of the chick's life are especially important as it is likely during this time that the maternal immunity is fading and new immunities produced by vaccines are building up. A freedom from challenge at this time can give the bird an assured future.

The danger of public nuisance from poultry

In recent years there have been increasing numbers of objections from residents near poultry sites who claim that they suffer nuisance from smell, dust, flies and noise. It is claimed that this is not only destructive to the quality of their lives but also causes ill-health. Though most of the complaints are more emotional than objective, everything should be done to reduce the possibilities of these nuisances and the following measures can help a farmer towards this end.

It is preferable for a mechanical ventilation system to have a positive wall extraction arrangement discharging the foul air downwards towards the ground. To further reduce the spread of smell and dust the extracted

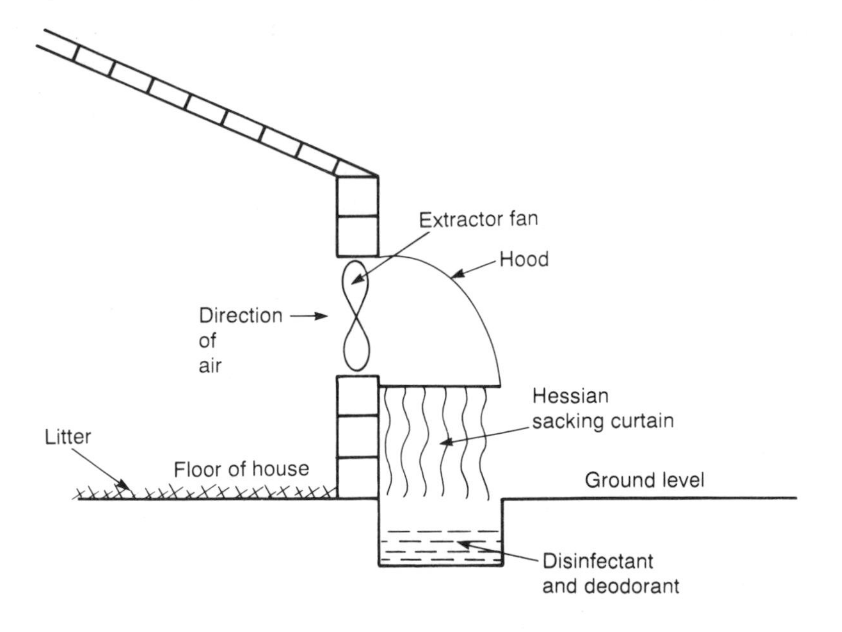

Fig. 8.1 Ventilation extraction to reduce dust and smells.

air can be discharged into a sump containing disinfectant and deodorant with a hessian curtain protecting the area of discharge (Fig. 8.1). Placing the fans on the walls rather than on the roof also reduces the possibility of noise which can be further reduced with sound insulation panels placed in the hood; in addition special purpose-made fan mountings will help further to reduce the noise.

When litter or muck are removed, the tractor should be covered once it is loaded before moving off the site, and the contents taken far away from any danger of causing smell to residents.

Good ventilation and environmental conditions will greatly help in reducing smell. One of the least pleasant occurrences is the smell that arises when there is extremely damp litter in the terminal stages of a broiler crop and when this is being cleared. Good temperatures in the house and ventilation, no over-stocking, proper choice and management of drinkers to reduce spillage and a sufficiency of good litter which is well managed will assist greatly. Many of the complaints that arise from nearby residents are magnified if such care is not taken.

One of the greatest helps in reducing the risk of smell is to plant a bank of trees between the poultry site and the houses. Psychologically it helps if the site cannot be seen but the trees also assist by trapping the dust and smell and also isolating the site itself from cross-winds.

The flies which may be associated with some sites will be due to lack of attention in the cleanliness of the houses and the surrounds, which thereby afford breeding areas for the flies. The whole of the site should be kept in the same good order as the houses and herbage cut or land cultivated. Flies may also breed in the dung in deep pit houses and this may need to be treated with insecticide.

The litter condition has also been improved by the addition of zeolites, which are crystalline hydrated aluminosilicates of alkaline earth cations. These have the ability to absorb and control the amount of ammonia and moisture and thereby reduce the nuisance but, in addition, they have beneficial effects on growth and egg production and reduce the amount of damage to the food pads and hocks of the birds. There are other additives to litter that make similar claims but they are not widely used due to their expense. Good management of the environment is above all the best way to reduce complaints of smell and manure.

Plate 10 A typical group of poultry houses – quite closely packed together and with very easy disease transmission between the buildings which are only about 5 m apart.

Plate 11 Interior of houses in *Plate 10*. The birds inside will be tightly packed, especially if they are in tiered cages. The photograph shows four tiers of cages giving a very heavy concentration of stock. The cages shown use belt cleaning daily.

Plate 12 With other cages the muck may accumulate in a deep-pit under the cages for up to about one year. Under these circumstances disease challenges are inevitable and protection from vaccines is most satisfactorily given by a special spray applicator as shown in *Plate 13*.

Plate 13 The Turbair Spray vaccinator.

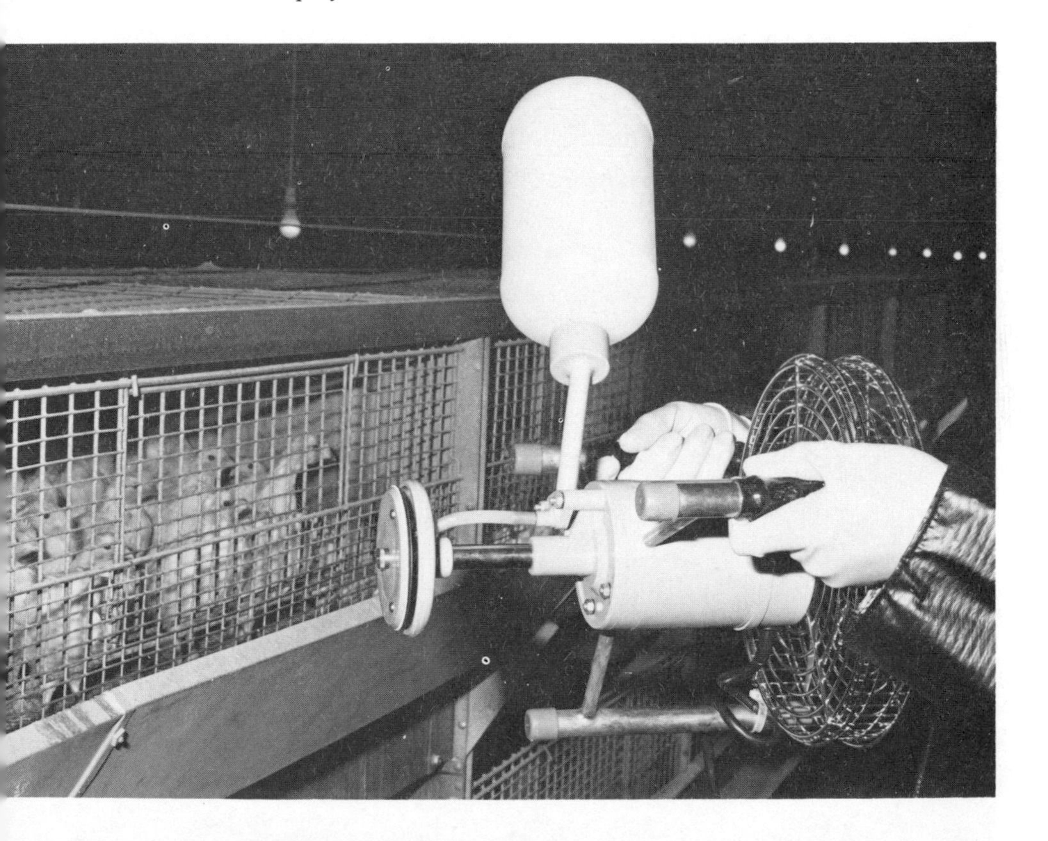

Plate 14 Hygiene and disease control are also helped by good roadways and surroundings, epitomised in this illustration of the end of a group of four broiler houses where access is confined to one end of the site, so reducing risks and disturbance to a minimum.

9 Systems of management

The rearing of birds

Floor housing

Chicks may be reared on litter on the floor from day-old until they are ready to enter the laying quarters. This is a popular and simple arrangement. There is no stress caused by changes in housing and resistance to parasitic diseases, such as coccidiosis, tends to build up gradually so the risk of disease from this type of source is reduced. Birds may be reared for any subsequent laying method in this way and it is also a popular practice to take breeding stock from day-old through to the end of the laying period in the same quarters. Some commercial layers are also reared in this way. It can be said to be extravagant on housing as the birds have to 'grow into' the space provided which must be sufficient for the adult. Those, however, who use single-stage systems (colloquially called 'day-old to death') are convinced that they gain more than they lose. The absence of early pressure on space does assist in healthy development with a minimum of checks to growth. Both challenge and resistance to disease are favourably influenced as all processes are gradual, and furthermore the labour involved in handling birds and cleansing and disinfecting the houses is reduced to a minimum.

With these systems of rearing, a floor area of 0.2 m^2 per bird should be allowed for the lighter breeds, and up to 0.24 m^2 for the heaviest birds – the maximum figure being required for broiler breeding stock. The floor area per bird can be reduced by about one-third by using housing of partly deep-litter and partly slats or wire. With this arrangement a useful proportion of the droppings fall into a pit under the slats or wire and the deep-litter remains in a much better condition. Also, by placing the feed and water points largely on the slatted area, the birds are not only encouraged to make more use of this area but water spillage goes safely into the droppings-pit.

In those cases where there are a combination of litter and slats, the young chicks can be brooded on the slats which are covered at first with paper bags and litter. After a few days the chicks are given access to the littered area. More often they are brooded on the litter and later, at a few weeks of age, they are trained to use the slats by the provision of a sloping run-up and the addition of water and food troughs to the raised area. Rearing on slats and wire alone can be used but is uncommon now as it presents husbandry difficulties such as feather-pecking, cannibalism and hyper-excitability. Wire with a mesh of 25 mm × 25 mm is satisfactory for all birds from six weeks of age, but if used earlier than this a cover of smaller mesh wire, approximately 12.5 mm is required.

Single-stage rearing is also carried out widely and successfully in special pullet-rearing battery cages, in which the birds are taken right through from day-old to placement in the laying cages. The usual system is to start by putting all the chicks in the top, or perhaps the centre tier, where the temperature can be kept to the required level and where management and observation are most easily carried out. They can then be spread, through all tiers within a few weeks of hatching, and indeed this change should on no account be delayed or the high density retained in one tier will cause serious problems of uneven and, or, retarded growth. Birds that are reared in cages should be housed subsequently in battery cages and not on deep-litter, straw-yard or free-range since they are likely to have a poor resistance to parasitic infection, having probably had no challenge at any time and hence building up no resistance.

Single-stage intensive systems require careful hygiene, nutrition and management in order to minimise the risk of vices and disease, especially as units may often contain many thousands of birds. They do, however, have the enormous advantages I have mentioned and simplify the whole process of management.

Cages are produced that are so suitably designed that chicks can be brooded in them and yet they may also serve as laying cages. They have, as in the floor systems that parallel this, advantages in favour of labour economy. In addition they tend to ensure that the chicks are in small groups, and also parasitic diseases are unlikely and the food and water are very accessible. Great care must be taken to choose good equipment that has all food and water points easily adjustable and, where nipple drinkers are provided, I have found it desirable, in my own work, to provide founts when the chicks first go into the cages. The floors may be of 20 mm wire netting or 25 mm × 12.5 mm, 16 gauge welded mesh. When the chicks are placed in these cages a temporary flat floor is used, later to be replaced by the sloping floor required for layers.

Multi-stage systems

Several rearing arrangements brood the chicks in a special brooding house and then carry them on afterwards, when the period of artificial heating is over, in various other systems of accommodation. A substantial number of these are traditional and often remain in use because the equipment is still available or those who have experience of them are so convinced of their virtue. For example, a traditional system is to rear the chicks in a tier-brooder which is suitable for the first three to five weeks of life. Tier-brooders have the advantage that they can be placed in many forms of adapted and non-specialist buildings but as room for expansion is limited they should not be kept in them for too long. The tier-brooder consists of two or three warmed brooder compartments largely enclosed and placed one on top of another, with runs with wire sides and top in front. The base is all of wire and droppings trays are provided underneath. A typical size of tier-brooder is 3 m × 1 m taking about 50 chicks to four weeks of age.

It is noteworthy that the tier-brooder and the building within which it is housed represent an area of great disease risk because of the quick succession of populations. It is therefore an essential of this system and indeed of any building where very young stock are kept in quick succession, that the building is depopulated after each batch and then cleaned, disinfected and fumigated. The risk is increased further by the high density of birds the brooder house can contain.

From tier-brooders the birds can really go to any other system: wire-floored carry-on cages, deep-litter, deep-litter and slats, straw-yard or free-range. However, it is always best to reduce the various forms of flooring used to a minimum since not only will this result in increased hazards from parasitic diseases, such as coccidiosis, but it will also cause a more prolonged check in growth. If tier-brooders are used a system which has much to commend it is to keep the chicks in tier-brooders for a maximum of six weeks and then put them straight into laying cages which are specially designed to cope with the feeding and modified flooring required for younger birds.

Outdoor rearing

A procedure that was once very common is to rear the chicks from hatching to about four weeks either in tier-brooders or on the floor below an ordinary radiant brooder. They are then transferred to a 'hay-box' brooder. This consists of a covered compartment 1 m × 1 m with a run approximately twice as long in front. No artificial heating is required because the covered section is packed round with hay or straw to conserve the chicks' heat and prevent draughts. Whilst this latter part is constructed of solid wooden sides and roof, the run has wire mesh sides and roof. The units are designed to be moved over the ground – which should be good pasture – but they are also sometimes used as fixed units mounted on straw bales.

The system is healthy and can produce hardy, vigorous stock at an economic running cost as no heat is used, but it has unfortunately serious faults in other ways, being labour intensive, and with high demands on capital, maintenance and above all stockmanship.

After the hay-box stage birds are normally reared in range shelters, or, indeed, they may be brought to this system from the other systems of management, especially if they have been reared up to then on the floor. Range shelters are a simple, easily made and light form of protection and consist of an ark approximately 3 m × 2 m with a height of 1.7 m at the ridge and 0.85 m at the eaves. A unit of this size will take 60 growers from eight weeks to about eighteen weeks. The roof can be built of metal, plywood, hardboard or felt and wire for extreme economy, the sides being of wire but a large overhang on the roof protects the stock from wind and rain. Alternative housing systems for birds reared on range are the Sussex night ark and the apex hut. The former is a hut of 2 m × 1 m with a slatted floor and solid walls and roof, whilst the apex hut is essentially similar to the other two, but is of triangular cross-section.

A few farmers prefer to rear their birds in smaller groups, confine them and protect them and yet still get the benefits of grass. In this case the fold unit is the best method. This consists of a hut measuring some 2 m × 1.3 m with a run extending 4 m in length in front. It will hold about 25 to 30 pullets to the point of lay. If fold units are moved frequently, which they should be to prevent a parasitic or other pathogenic build-up of organisms in the soil, the droppings will be spread uniformly over the ground.

Broiler chicken management

The systems used for rearing broilers are probably more standardised than any other arrangement. Almost invariably the birds are reared from day-old to about 46–70 days in a controlled environment house on built-up litter of wood shavings or straw or a mixture of the two types. Because the birds grow so quickly any upset in the condition of the litter, which must be soft and friable, will be very likely to lead to a damaged carcase due to bruising of the breast muscles which are the most valuable part of the meat.

A typical broiler house consists of a building of 10000 to 20000 capacity, rectangular in shape, and from 14 m to 28 m wide. Birds are reared commercially to give a maximum live weight at the end of the growing period of 34 kg/m^2. The lightest weights of bird are marketed at about 1.4 kg live weight, though the more typical weight now sought by the market is a 1.8–2.0 kg bird. Details of the various techniques of brooding, feeding, drinking, litter management, lighting and disease control are given in the relevant sections of this book. It is of much interest to the industry that after the culmination of many years' work in various centres in the United States and Europe it now seems likely that satisfactory methods are available to rear broilers in cages on suitable flooring of perforated plastic form, though such systems have not yet been very widely used. The absence of the commercial adoption of cage rearing appears to be due not so much to technical difficulties as to the enormous capital requirements involved in providing new houses and equipment. Its advantages would be several, perhaps the greatest being the mechanical handling of birds in their rearing cages to the processing plants, so lessening labour, stress and damage.

Laying systems

Traditional free range

Only a few layers are kept under the traditional free range system, most now being housed under the modified arrangements described in Chapter 17. With traditional free range systems housing is simple and cheap but any real environmental control is impossible. The birds may be housed in a movable colony house of 3.3 m × 1.7 m on skids or wheels holding up to 50 birds. This type of hut has a solid wooden base and is fitted with perches, droppings board and a nest. An alternative is the slatted floor house which is similar but does allow a heavier concentration of stock, a hut of 2.7 m × 2 m holds 60 birds comfortably. It has skids or wheels and is fitted with nests, a broody coop, troughs and drinkers.

Free range systems can be healthy but it is entirely wrong to believe they are inherently healthier than indoor systems, and they have serious disadvantages in that the birds and the stockmen are at the mercy of the elements and winter production can be particularly poor. Some of these disadvantages may be overcome by keeping the birds in a fold unit, as the environmental control is somewhat better. The fold unit consists of a covered-in roost with a slatted floor, measuring about 1.7 m × 1.7 m, with a height of 1 m and with a totally wire enclosed run in front of about 4 m long. The whole unit is sturdily built of timber to withstand daily movements on to fresh ground.

The fold unit has a capacity for 25–35 birds, depending on the type of bird housed. Whilst the birds do benefit from the advantages inherent in keeping stock in small groups, and in the movement of them to fresh clean ground each day, the cost is out of proportion to the return.

In both these arrangements production can cease altogether under very severe wintry conditions and it is difficult to see how such costly methods can return to popularity.

Semi-intensive – fixed pen with outside run

The fixed house or 'cabin', with single, two or more alternative runs, is a system that is only used to any extent now by the domestic poultry keeper. A typical unit consits of 50 birds kept in a house 3 m × 3 m with two runs each 10 m × 10 m, the birds alternating between them at six monthly intervals. The cost of such a system is high both in capital and labour, while the climatic control over the birds is poor. Even when alternating between two runs a certain build-up of disease may occur. Eggs from this system are often dirty. It ranks as being perhaps the worst system of all in theory and in practice but can function in the hands of the domestic poultry keeper where the normal economics of the market place do not operate.

Fully intensive housing systems

Deep-litter systems

One of the first really successful methods of housing layers intensively was the built-up or deep-litter system. A detailed description of the importance of good litter management in which great care must be taken was given earlier (chapter 2). After some forty years of use it remains as a thoroughly satisfactory system though of rather high capital cost. Now, because of this, commercial egg layers are not kept on deep-litter but breeders and broilers are. The litter is either wood shavings about 350 mm deep, or increasingly popular is the use of chopped straw; this may be mixed with or replaced by peat moss or sometimes paper shavings but these are rather poor substitutes.

In the earliest designs of deep-litter houses movable perches, drinkers and feeders were used so that, as far as possible, droppings and water splashings were distributed as evenly as possible by moving these around the house. However, under this system, birds require about 0.27 m^2 to 0.36 m^2 of floor space per bird if the litter is to work properly. It is possibly a cheaper arrangement to have a slatted area over a droppings pit with perches and possibly food and water troughs situated here as well. In this way the litter receives proportinally less droppings so the density of stocking can be increased. If the nest boxes are suitably disposed, with feeding and egg collection by hand, the two operations can be combined in a single movement round the house. Automatic feeding is more usual and there are now several systems for satisfactory automatic egg collection, so the 'chores' can be quite well reduced.

With an area of deep-litter combined with a droppings-pit of roughly equal proportion, 0.18 m^2 per bird is sufficient for 'heavies' and 0.14 m^2 is enough for the 'lights'. In more recent years it has become quite popular to keep breeding stock on deep-litter without a droppings-pit but sometimes with whole house heating provided from an oil- or gas-fired hot-air heater or radiant gas heater. This keeps the litter in good condition even in winter. But even without the heater, provided there is very good insulation (see later section), the absence of a droppings-pit can be tolerated. By eliminating the droppings-pit, there is the advantage that the capital cost is lowered and management, particularly cleaning, is made easier. It also helps to make ventilation better without the obstruction of the pit and it promotes the better distribution of the birds on the floor since there is a general uniformity of the conditions, and no reason for areas to be favoured. Because the cost of fuel is high it emphasises even more clearly that the key to the correct temperature needed in a deep-litter house is good insulation rather than artificial heating.

The straw-yard

The hen yard is in essence a much cheaper version of the deep-litter house, using a thick bedding of straw only as the litter and a different approach to environmental control. It was introduced in its original form in the 1950s as a cheaper system suitable for the arable farmer who had abundant straw and a selection of old and otherwise redundant buildings. Because it included a large uncovered yard area at that period of its development it failed to compete with fully intensive systems since it gave lower yields and often dirty eggs. The disadvantages had to be balanced against the very low capital cost of the system, assuming some cover was already available. One of the worst disease problems that occurred was an infection with intestinal worms, the viability of the worm eggs being favoured by the damp conditions, though if plenty of straw was used and good drainage arranged, the danger could be somewhat reduced. Figures have been produced to show that as good a profit may be obtained per

Table 9.1 Egg production under different housing systems

	Eggs/bird (dozen)	Food/bird (kg)	Food conversion
Free range	14	54	11.4
Intensive (poor control)	17	50	8.6
Intensive (fully controlled environment	20	43	6.3

Table 9.2 Stocking densities

		Floor space/bird (m^2)
Layers		
Deep-litter		0.27
Deep-litter and slats in ratio 2:1		0.18
Deep-litter and slats in ratio 1:2		0.14
All slats or welded mesh		0.09
Multi-bird batteries and house		0.18–0.27
Totally covered straw-yard		0.27
Rearing		
Tier brooders		0.02 at 3 weeks
Floor rearing		0.09–0.14
Hay-box brooder		0.03 to 8 weeks
Sussex night ark	plus pasture	0.04 to point of lay
Range shelter	plus pasture	0.06 to point of lay
Fold units	plus pasture	0.06 to point of lay

bird from this system as from some of the more fashionable ones but the arrangement requires a high degree of skill in management and has no widespread application at present.

The construction of the uncovered straw-yard is simple. The only essential is a shed or yard – uninsulated and open-fronted. In front of this is built an uncovered compound. The whole area is provided with deep straw litter. Each bird is usually given up to 0.18 m^2 of covered space and about twice as much uncovered yard area. Some drinkers and feeders are placed in the uncovered area whilst the remainder of the food and water troughs are placed in the covered area. It is preferable if the floor of the latter is entirely of wire or slats on which may be placed sufficient feeders and drinkers, to provide for all the birds in severe weather, but an alternative is to have two-thirds slats and one-third litter. If the nest boxes are placed so that the birds have to walk across the slats to enter them, cleaner eggs are produced, so helping to overcome one of the main drawbacks of the hen yard, which is the production of dirty eggs in wet weather.

As the system is intended primarily to make use of existing open-fronted yards, cart sheds and so on, the problem of avoiding draughts and cold winds must be dealt with. Both the yard and the house benefit from a solid front to a height of 0.6 m to 1.0 m, a cheap way of providing this being to use corrugated sheets placed horizontally. It will depend on the height of the eaves and the overhang, if any, of the roof as to whether further protection is necessary. If the eaves are no more than 2 m and the overhang 1 m or more no further protection may be required. Usually, however, these ideal requirements are not satisfied, in which case a simple front may be constructed of straw bales and polythene. A reasonably free circulation of air must nevertheless be allowed in the house, otherwise damp litter and condensation problems may result.

At Cambridge we have used for over twenty years an unsophisticated form of straw-yard which is totally covered but uninsulated and depends on natural ventilation. This is fully described in chapter 17.

The slatted floor system

Some years ago an arrangement was developed for keeping layers on totally slatted or totally wire floored housing. It is the most concentrated method of housing layers on the floor at a stocking rate of 0.09 m^2 per bird. Labour requirements are low, housing costs can be competitive but apart from the continuing enthusiasm by a few poultry farmers who made it work, and still do, in general it has been a failure with an excessive number of floor eggs, which are lost. Extreme behavioural problems were also created, probably by the boredom inherent in a system such as this allied to the very close proximity of a large number of birds. Nevertheless in some areas of the world it is still a very popular form of housing, especially in the north-east of the U.S.A. which is an area climatically not so different from the U.K.

Layout for floor laying houses

The most important point about the disposal of the equipment in a floor laying house is to arrange it so that the various tasks associated with management – feeding and egg collection being the major ones – can be carried out easily. In a small building holding up to 500 layers, a service room may be placed at one end of the house and the nests arranged so that collection is from this room, the nests forming the upper part of the dividing wall between the house and the service room. In larger buildings with automatic feeding the more usual arrangement is to have a central service passage and nests on each side so that a quick collection is easily made. If the birds are manually fed it is preferable to have the nests against the wall and an overhead monorail and conveyor adjoining the nests. The conveyor can then be used both to collect the eggs and to take the food to the tubular or trough feeders which are placed on the opposite side towards the centre of the house. When this design is used in a deep-litter house, a central droppings-pits is installed and the majority of the waterers are placed over this to minimise the harmful effects of splashing. It is preferable if the birds walk across the slats before they enter the nest box, for the same reason as in the straw-yard, that it tends to keep the feet cleaner and the egg is less likely to be soiled. With such an arrangement in a wide span building, a centre passage would be used with nest boxes on each side and droppings-pits immediately adjoining. There are, however, those who prefer to site the droppings-pit at the side of the house for ease of construction, the usual pattern being to have it on one side and a central deep-litter area and nests adjoining a passage on the other side. This passageway will also take the food trolley where hand feeding is practised. The house will usually be cleaned out annually and to assist this the slats, perches and divisions should be easily demountable and the end doors should be sufficiently large to allow tractor entry. Mechanical methods of daily cleaning under the droppings-pit with a conveyor belt or scraper are used but have not proved popular owing to the high capital cost. In fact the whole trend in deep-litter housing is simplicity, with the omission of any droppings-pit and the use of automation in egg collection, to leave the operator as a skilled overseer of the health and well-being of the birds.

Laying cages

The system of keeping birds in cages, often three or four tiers high, exceeds all other housing systems in popularity. This is due to several factors. It has been due partially to the breeding of small birds which can be kept housed much more densely than hitherto and also due to the fact that whereas one bird per laying cage was the normal practice in the original forms of cage, it is now the practice to keep 2–25 birds per cage at a greatly reduced cost compared with all other systems, apart from the

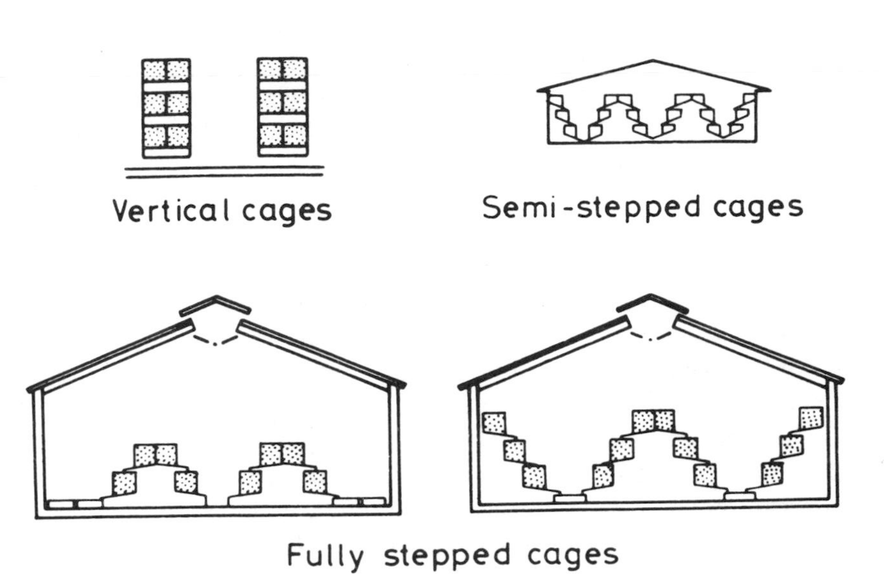

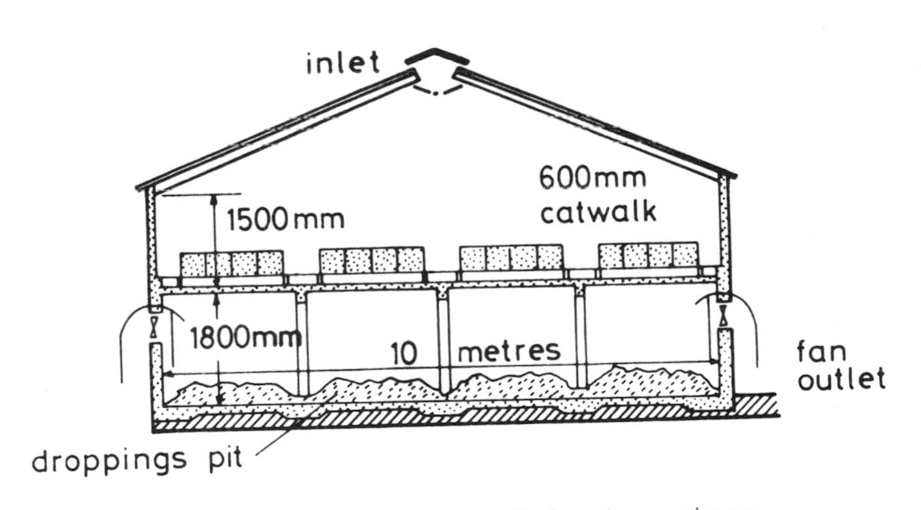

Fig. 9.1 Cage systems.

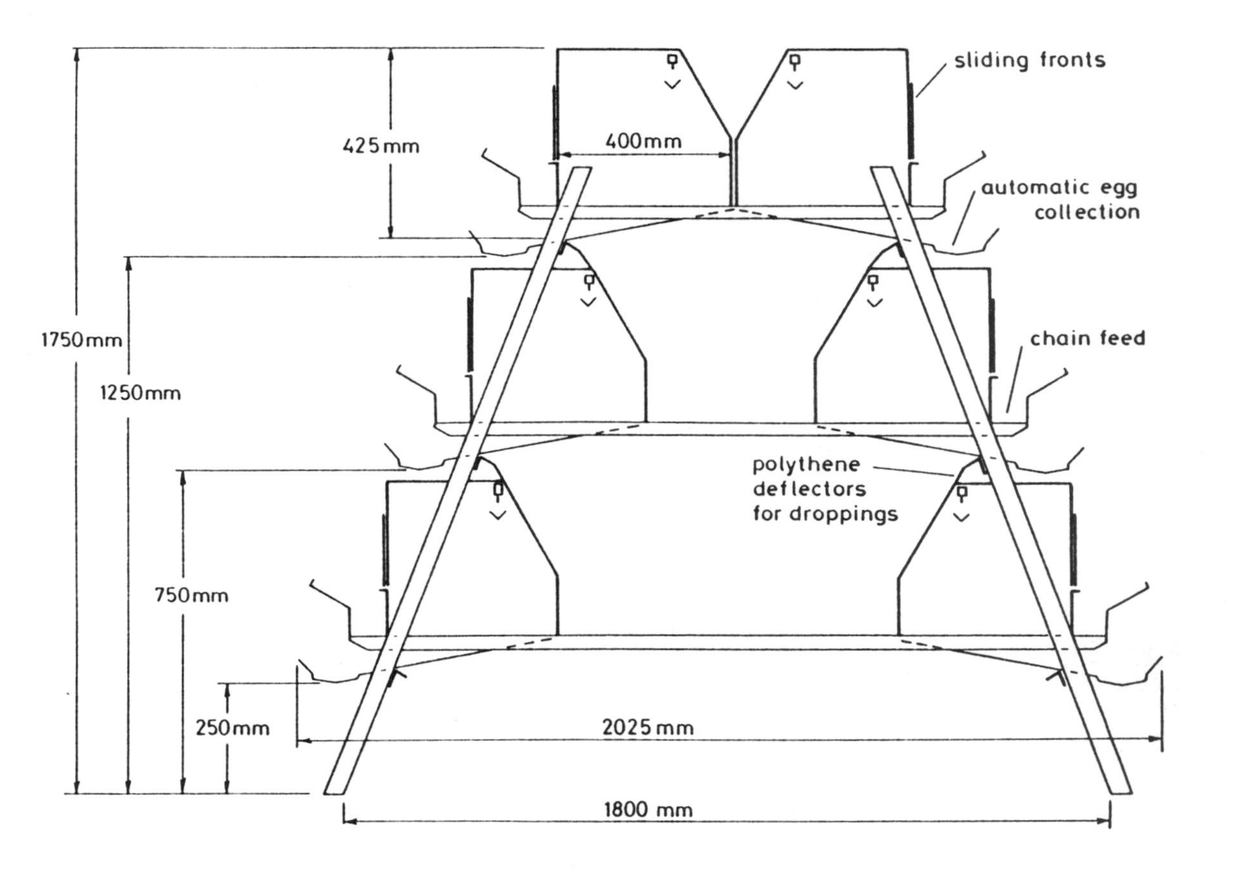

Fig. 9.2 Semi-stepped cages – great economy in cost.

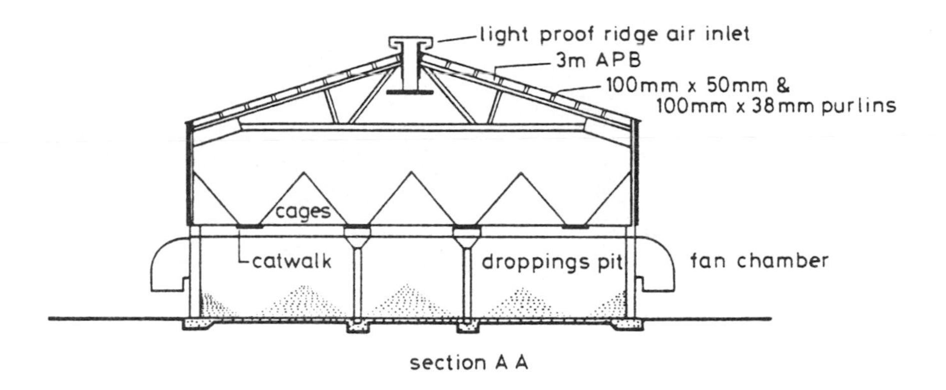

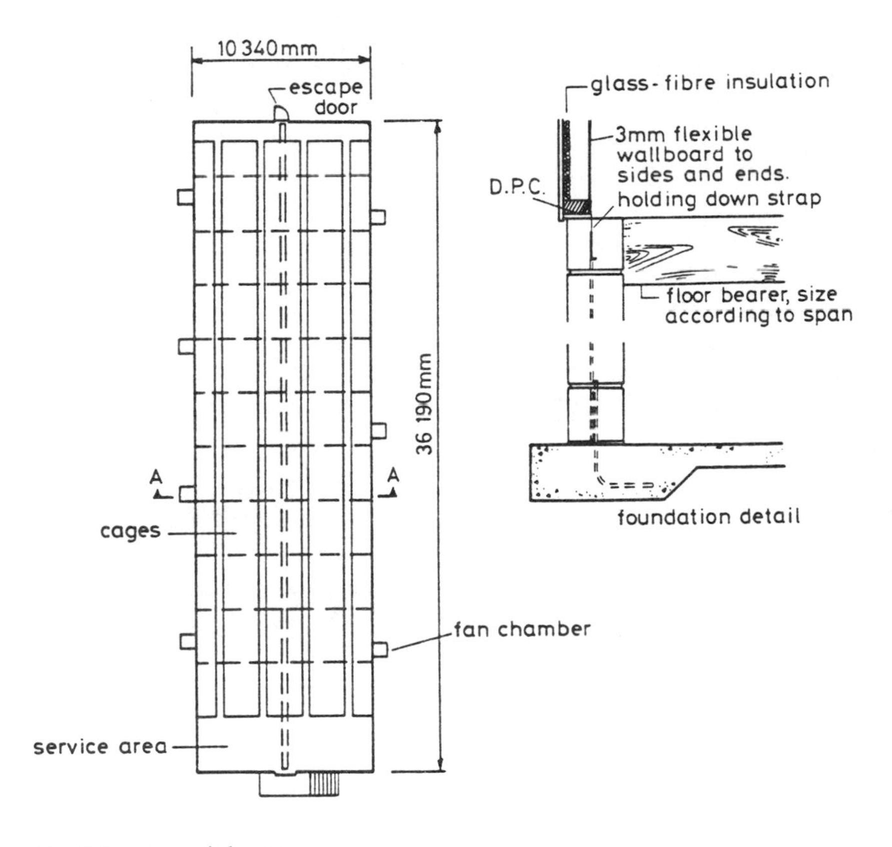

Fig. 9.3 Deep-pit house.

straw-yard. The most popular cage systems house 3–5 birds in a cage. With a multi-bird cage system the *total* floor area per bird can be no more than 0.06 m^2. Cages vary in width as follows: 225 mm to 300 mm for a single light hybrid; 300 mm to 325 mm for a single heavy hybrid or two light hybrids; 350 mm to 375 mm for two heavier hybrids or three light hybrids; and 525 mm for five light hybrids or four heavy hybrids. Cages are usually 450 mm deep and 450 mm high at the front, sloping to 350 mm high at the back. The floor extends at least 150 mm in front to form an egg cradle and the droppings-tray or belt, if there is one or the other, has a minimum clearance of 100 mm below the floor of the cage. Each bird should have 100 mm of trough space.

The various cage systems

There are several different types of cages that can be used. The three major groups are as follows: *stacked cages*, one on top of another are three or four tiers high and are mostly mechanically cleaned by a scraper, a plastic belt or occasionally by using disposable paper belts. Some are still hand cleaned. Occasionally, more than four tiers are used, even up to six or seven, and an intermediate catwalk is required to service the top tiers.

The next group of cages are the *Californian or stair-step*, often called the *deep pit system* because of the way it is installed. These cages are staggered, as in Fig. 9.3, so the droppings go into a large pit under the cages and can build-up right through the laying cycle when they are cleaned out by a tractor and a fore-loader. These also may be two or more tiers high, three being usual and four now becoming popular. A recent form of stair-step is the 'semi' form in which the upper cages are set partly over the lower ones, thus economising on space.

Finally, there is the *flat-deck system* which is of single tier cages usually in 2 m wide blocks consisting of four cages with a double line of automatic feeders and belt egg collectors running between the inner and outer cages. This is the most expensive system but the labour cost can be very low and the mechanical installation is quite simple. Cleaning can be either by a scraper system over a shallow pit or the droppings can be in the usual deep-pit. With fully automated flat-deck cages it is possible for one man to handle up to 20 000 birds, whereas other systems are likely to do considerably less than this.

Stacked cages used to be the most popular; probably the main inherent difficulty with these units is the cleaning system whether it be by belt or scraper since the 'runs' should not be much longer than about 30 m or the mechanical equipment can be troublesome. Also with travelling food hoppers and belt egg collection there can be the same problems with the equipment if it is expected to do too much. Manufacturers tend to be too optimistic about the robustness of their equipment under the extremely exacting conditions of the poultry house. It has been my experience that this is one of the major problem areas since (a) when equipment breaks down it may be extremely difficult to get it repaired, and (b) the attention needed to repair the equipment may cause considerable disturbance to the birds and so affects production. In the period between breakdown and repair the servicing of the birds may be extremely difficult, if not impossible.

Adequate passageways between rows of tiered or stacked cages are very important for ease of all procedures. If a unit is a hand-cleaning one, then 1 m is advisable, but if this is not possible 850 mm will suffice. Passageways of 600 mm are general with systems where there is automatic egg collection. Whilst this is in theory adequate, and in practice sufficient with flat-deck cages, it is unwise to have birds so close if it can be avoided. In any system of cage disposition the passageways serve to some extent as the means of distributing the air to the birds and wider passageways contribute to the maintenance of a uniform, draught-free environment. Conversely, if the passageways are very narrow the closeness of the birds will contribute to the build-up and spread of respiratory and other diseases.

Feeding

The simple system of filling the hopper by feed from a barrow is perfectly satisfactory for the small unit but the usual arrangement is for a continuous trough filled by travelling hoppers which are either moved by hand or mechanically to fill the tiers simultaneously. The feed hoppers themselves can be filled automatically from an overhead auger operating from a bulk food bin. An alternative arrangement sometimes used is to run a chain feeder round the cages, and a further system is to use a barrow fitted with a chain and flight lift which puts the food directly into the trough. My advice, which I cannot overemphasise, is that any system should be kept as simple as it can and it must be designed so the chances of a fault are as remote as possible.

Drinkers

There are three arrangements for providing water to battery birds, i.e. continuous troughs, cups, or nipples.

With the continuous trough, served by a drip feed or tippler tank at the end, there is the advantage that the birds can see water, have it always in front of them and can physically dip and wash their beaks in the water. This system has the disadvantage that the troughs should be cleaned regularly and there is some food wastage. Much more popular now are valve or 'nipple' drinkers, which are valves fitted into plastic piping which the bird can depress to produce a modest flow of water. These are simple, clean and cheap. One nipple should be provided for each 4–5 birds. The best arrangement is to place the nipples by the partition between cages so that birds from both sides have access. Thus all the birds in a cage have access to two nipples and are unlikely to be denied sufficient water.

The cup drinker is a small cup about 20–50 mm in diameter filled by a valve which the bird operates automatically with drinking. It is a good compromise between the two systems being clean, hygienic, providing a 'dip' for the beak. There is some evidence that birds do rather better with water in front of them but it is not consistent and nipples cannot be condemned on the grounds of this rather flimsy evidence.

The 'get-away' cage

A mention should be made of the 'get-away' cages as these have been designed for an important reason. The cage (Fig. 9.4) provides three areas – a horizontal wire floor, a nest box and areas for perching. They also, it will be noted, have two levels of feeders and drinkers. These are to provide an environment which is more akin to the 'natural' state insofar as there are perches for resting (the 'get-away' part), a separate area for laying and extra facilities for feeding and drinking. In this less stressful environment it is felt there may be less likelihood of a welfare problem and better production, so compensating for the extra cost. The procedure is under study and it is too early yet to give definite answers as it has not been widely adopted commercially, but such results as are available show a slightly poorer production and dirtier eggs. The advantages remain open to question.

Record keeping

It is always essential to record carefully the results of performance in all systems of poultry production and from which the margin of profit (or loss) can be evaluated. For example, with a laying flock one would take the following records at least:

1. The total number of birds housed.
2. The cost of the birds or rearing costs if done by owner.

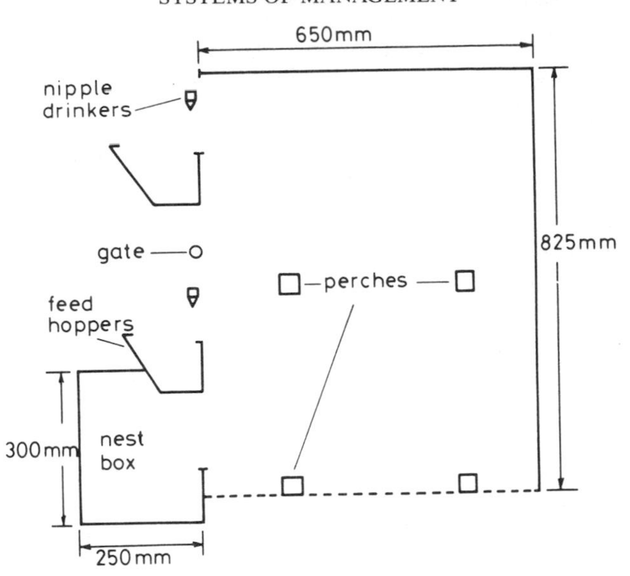

Fig. 9.4 Get-away cage.

3. Eggs collected daily.
4. Food consumed daily.
5. Egg sales.
6. Grading details.
7. Mortality and culls daily.
8. Birds sold at end of lay and their value.
9. Labour costs.

Most useful calculations for a farmer to make quite frequently – usually weekly – from the figures collected are as follows:

$$\text{\% egg production per week} = \frac{\text{Total eggs per week} \times 100}{\text{Average no. of birds per week} \times 7}$$

and

$$\text{\% egg production based on number of birds housed (hen housed)} = \frac{\text{Total eggs to date} \times 100}{\text{No. of birds housed} \times \text{no. of days since housing}}$$

The latter gives the percentage of production based on the birds housed (hen housed average) and is a frequently used figure. A graph of production day by day is invaluable to give a quick check on productivity and is a clear way to detect trends early on. The profit and loss of the farm from time to time can be worked out quite readily from these basic figures.

For broilers a popular yardstick of efficiency is the European Production Efficiency Factor (EPEF), which is as follows:

$$\text{EPEF} = \frac{\text{Av. liveweight} \times \dfrac{\text{Total liveweight sold}}{\text{No. of chicks started}}}{\text{Av. age days} \times \dfrac{\text{Total feed used}}{\text{No. of broilers sold}}} \times 10\,000$$

All measurements are in kilograms (kg). If this seems rather complex, the most important needs with broilers which can easily be calculated are:

1. *Liveability* – the percentage of chicks alive at the end of the crop.
2. *Feed Conversion Efficiency (or ratio) FCR* – divide the total weight of feed used by the total liveweight started.
3. *Feed cost per kg liveweight* – multiply FCR by cost of feed per kg.
4. *Average weight per bird* – divide the weight of birds sold by the number of birds sold.
5. *Gross margin per unit of floor space* – subtract the feed cost from the gross income and then divide the result by the floor area (m^2 or ft^2).

To aid the poultry farmer in keeping records, excellent assistance is usually available from national compounders, breeders, hatchery organisations or the government advisory services. The poultryman is exhorted to make good use of these invaluable facilities as they remove the tedium and make record keeping a useful and fascinating job.

Plate 15 Interior of deep-litter poultry houses containing heavy 'broiler breeders', normally housed at about three birds per square metre. The automatic chain feeder is in the foreground whilst the self-filling drinkers can be seen at several positions. The individual laying boxes are visible on the partition between the pens.

The management of broiler breeders is one of the most demanding undertakings for the poultryman. A good feeding routine, excellent management of the litter, fine control of the environment and constant and careful care and handling of the birds are necessary to achieve satisfactory results. Good management can give up to 140 fertile eggs per bird but under poorer control the number may be barely 100.

Plate 16 Exterior of the deep-litter breeding house shown in *Plate 15*. The building has reversed air flow, with ridge air intake and side wall extraction. The bulk food bins are a prominent feature but the overall appearance is attractive with the light coloured fibre-cement roof contrasting with the creosoted timber exterior. The wall fan cowls are fitted with special devices to reduce back-draught.

With sites such as this, great importance is attached to the general standards of cleanliness and hygiene around the buildings since they will contribute to the overall success of the operation.

10 Breeding and hatching

In other sections of this book I have emphasised many of the essentials for the production of fertile hatching eggs. The sections include those on nutrition, housing and disease control. It must be emphasised that the effective demand on nutritional elements by breeders is especially great. In recent years the production of breeding stock has been markedly improved so that with the feeding of restricted rations it becomes more important than ever that the essential vitamins, minerals and amino acids are supplied in adequate amount. This will have a profound effect on the fertility and hatchability results. Special attention should be given to the nutrition during the depths of winter and early spring when the quality of the eggs and the chicks can often deteriorate. As this tends to be a constant seasonal problem it is difficult to understand why this is so often tolerated. Special attention should be paid to vitamins A, D_3, B_{12} and the B vitamins in general, and to manganese, zinc and choline.

It is also worthy of reiteration that healthy stock is particularly important for the production of sound hatching eggs. All diseases have an adverse effect on the ultimate viability of the bird, the eggs produced by it and the chick which may emerge. Some diseases in particular, such as infectious bronchitis and EDS '76 have an almost catastrophic effect, always for some time and often for the remainder of the bird's life (see chapter 11).

The fertility of poultry is partly dependent on the genetics and selection of the bird as inherent qualities, but is rather more dependent on good management than inheritance. A very important feature is the mating ratio – usually the optimum ratio is 6–7 males to 100 females for pullet replacement and 7–9 males per 100 females for broiler breeders. It should be borne in mind that once a group of males are 'made-up' they themselves cannot usually be interfered with.

Hatchability as well as fertility is affected by genetic factors but once again it is management which transcends genetics in its effect.

It is vital that hygienic standards in the breeder house are impeccable to avoid infections entering the incubator either within or on the shell of the egg. Nests must be clean and the litter dry; only clean wood shavings or other sterile litter should be used, and certainly never damp hay or straw. It must be changed regularly and eggs laid on the litter should not go into the incubator if at all possible. The greatest danger is that the hatching eggs will be contaminated with *Aspergillus fumigatus* (see chapter 11) which, if it gets into the hatchery, may be very difficult to eliminate.

For the best results, hatching eggs should be collected at least three times a day – in this way they keep cleaner and can be cooled more quickly. Selection of eggs subsequently is important, removing all sub-standard eggs with mis-shapen, cracked or thin shells. Eggs which are either too big or too small are undesirable and usually those between 48 and 64 grams (1¾ to 2¼ oz) are favoured. As soon as possible after collection the eggs should be fumigated with formaldehyde in a special cabinet, though they may be dry cleaned, washed or dipped in specially sanitised solutions at a temperature slightly warmer than the eggs. Great care must be taken that the sanitising material is safe and is frequently changed. It is not uncommon for dirty solutions to do more harm than good by spreading infection.

The conditions under which the eggs are stored before incubation need good control. The optimal temperature range is 13–16°C (55–60°F) with a high relative humidity of 75–80 per cent. In order to maintain such conditions, all farm egg stores should be well insulated and preferably equipped with an appropriate air conditioner if the eggs are to be stored for long. Pre-incubation storage should not go on for more than 7 days but a limit of 4 days is preferable since up to this limit there is almost no loss in hatchability if storage conditions are good.

The actual handling of the egg needs great care at every stage – and this does not apply just to the hygienic measures. The living embryo that it contains is marvellously protected but not so much so that handling cannot be abusive. Gentle transfer of the egge from one place to another is required and will subsequently improve the hatchability. The use of plastic setting trays has much to commend it as the trays can be used for on-farm traying so that handling subsequent to this is eliminated and fumigation is more effectively achieved. Plastic trays can be properly cleaned and sterilised at the hatchery and then re-used, which the old-fashioned fibre trays could not.

In the course of incubation eggs are 'candled', that is a bright light source is placed under the eggs and on viewing from above those eggs that are fertile appear relatively opaque due to the opacity of the developing embryo, whilst the non-fertile ones are clear. The 'clears' should be removed without delay as they can be a source of disease if they remain in the incubator and break.

Incubation

There are two principal types of incubator – the smaller cabinet incubator or the large walk-in mammoth incubator. These are usually divided into two processes, the 'setters' which receive the eggs for the first 18 days, subsequently entering the 'hatcher' for the final 3 days. A few eggs are, however, still hatched in small 'flat' incubators which are usually combined setters and hatchers. These, in which the eggs lie flat on their sides, are suitable for hatching small numbers of eggs.

Conditions required within the larger incubators are as follows:

	Temperature	Relative humidity
1–18 days	37.5–37.8°C (99.5–100°F)	60%
18–21 days	36.9–37.5°C (98.5–99.5°F)	60%

The conditions required in the small flat incubator without mechanical aids to the ventilation are rather different – a temperature of 38°C (103°F) is needed to ensure proper conditions in the centre of the egg.

Any wide variation from the optimal figures is likely to lead to serious consequences. For example, if the humidity is too low, excessive moisture is lost from the egg and many chicks will fail to hatch. If the humidity is too high, the chicks may hatch early and are excessively wet.

In the larger incubators the eggs are placed in trays with the broad end higher than the narrow end. The eggs must be turned frequently to ensure that the embroys does not settle and adhere to other structures. Turning is usually done automatically hourly through 45 degrees to the vertical. No turning is necessary in the hatching compartment.

Control of the ventilation by mechanical means is always carried out in large incubators so that the critical needs of the eggs for supplying oxygen and removing carbon dioxide are achieved. Carbon dioxide in particular, must not go above 0.4 per cent and oxygen must be at least 17.5 per cent. Incubator manufacturers do design all their machines to achieve the required conditions in various ways and their directions should be followed for the best results.

Most incubators are powered entirely by electricity and if there is a power failure high levels of carbon dioxide could quickly kill the embryos. Because of this, a stand-by generator is an essential for the immediate takeover of power when the mains fail.

The precise design of the incubators – setters and hatchers – varies according to its manufacturer and the whole process of design and management of hatcheries are becoming increasingly sophisticated and complex and are clearly outside the scope of this book. However, since the modern hatchery represents the greatest area of disease risk in a modern poultry enterprise, being a bottle-neck through which every chick passes – and by its usual large size being at even greater risk – we should carefully consider some of the essentials which have a bearing on these facets and which have not already been touched upon.

The incubators must be so ventilated that each unit receives fresh air from the outside and discharges its stale vitiated air to the outside, thus ensuring a minimum of air pollution in the hatchery itself. All surfaces of the building and the incubators must be of the best materials to ensure that they can be sterilised readily and efficiently. In many modern installations it is appropriate to actually filter the air into the incubator machinery. Within the hatchery there must be a 'flow of air' such that it can never pass back from 'dirty' areas, such as the hatchers, to clean areas and the setters and egg holding rooms. In the same way the movement of people and equipment has to be carefully designed and managed to give the same clean flow principle. Figure 10.1 shows the complexity of the modern hatchery and the various essential parts. Eggs will be received from the farms into the egg reception area, which should be big enough to take about a day's supply; if the eggs are to be held longer they must go after fumigation into a special egg holding room. Fumigation will take place again before the eggs go into the setter. The fumigation chamber should be sealable when fumigation is taking place and have powered ventilation to disperse the gas after fumigation.

It is especially important that where there is a separate hatching area – usual where mammoth outfits are concerned – that is well sealed from the rest of the hatchery, the discharged air which will contain a good deal of potentially polluted material should be collected by traps and, or, filters so that there is a minimum chance of recirculation of the extracted air back to the input side.

In the day-to-day management of the hatchery it is very useful to aerosol the atmosphere frequently with disinfectants. Care must also be taken with the staff who have to practise impeccable personal hygiene and should, from time to time, be checked for salmonella infections.

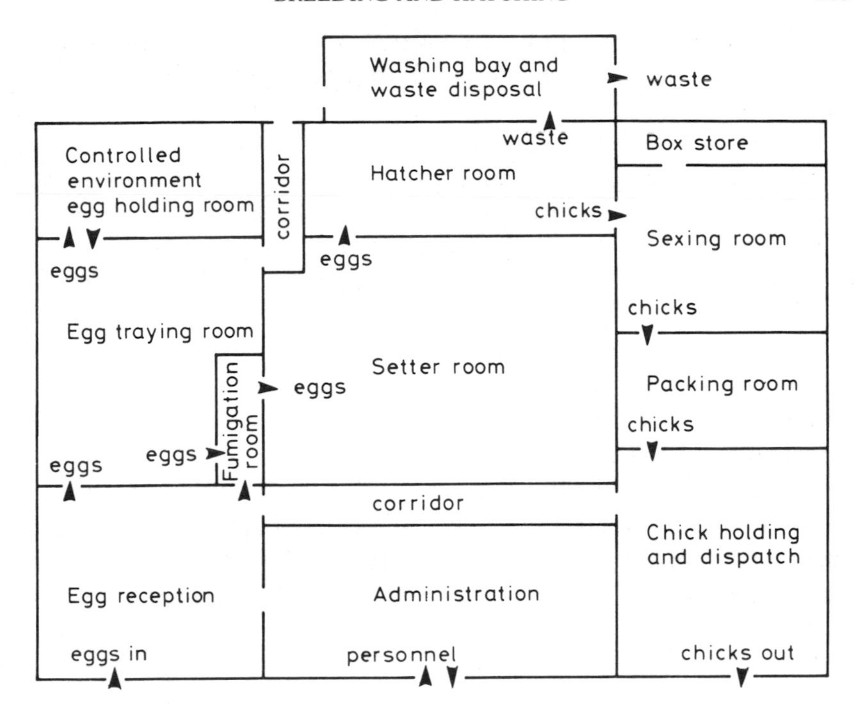

Fig. 10.1 Specimen hatchery layout (not to scale – diagrammatic only).

Plate 17 A group of three thermostats giving variable speed operation of fans. There are a number of devices that achieve such control but in essence the thermostat (or thermistor) devices in the house are wired to control boxes of some complexity.

Plate 18 A typical control box which incorporates the 'brains' to regulate heating, lighting and ventilation.

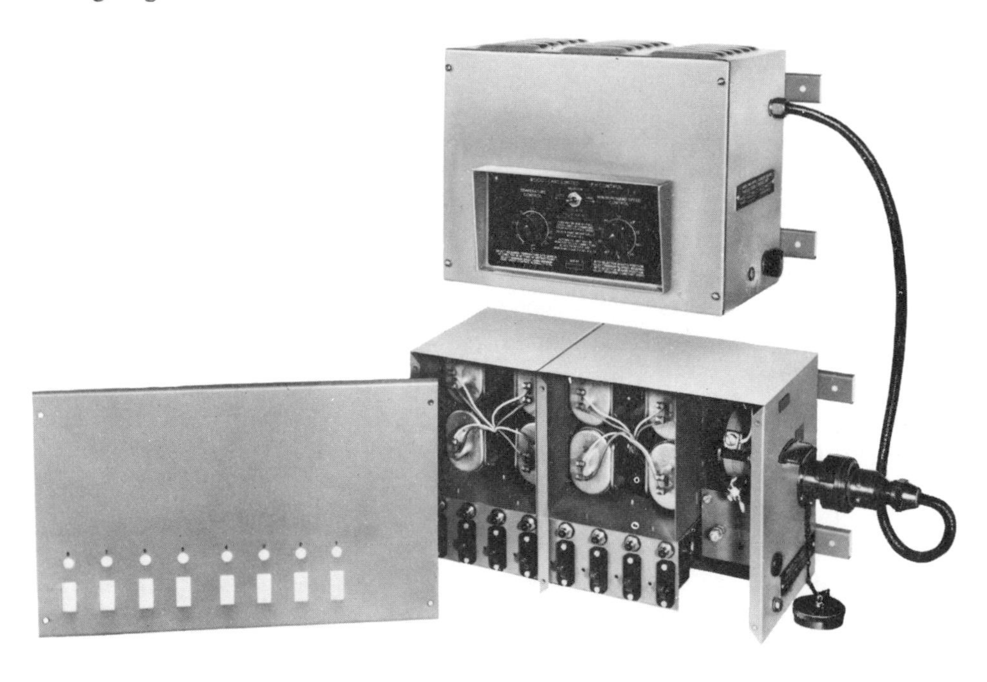

11 The health of poultry

This section of the book describes the diseases of poultry, their signs, prevention and treatment. It is in no way intended to be a catalogue or dictionary of diseases to be available only for reference on those occasions when it appears to be needed – usually, incidentally, after a disease has 'struck' and it is already far too late to control it effectively! Rather the purpose of this part of the book is to explain the circumstances and conditions under which specified diseases are liable to occur and how, by good management built on the foundation of basically sound units and buildings, not only will disease be far less likely to occur but, in addition, productivity can be enhanced at all ages and stages.

The most useful way for the poultry farmer or attendant to consider the incidence of disease is on a life-cycle sequence and this will be done by starting with the chick and tracing the likely disease problems through life.

The chick

In earlier sections of the book the management of the day-old chick has been stressed in some detail. There is no escaping the fact that the day-old chick is peculiarly vulnerable to stress, and conditions must be accurately maintained to avoid either chilling or overheating. Unfortunately, as birds are kept in larger and larger numbers in one house, unless very sophisticated environmental control systems are used it is almost impossible to provide a uniform environment and whilst the conditions required by the chick are well-known, the ability of the housing to provide them properly is often lacking. Thus there may be almost immediate losses from chilling in certain areas of a house, but in general the losses in the chicks in the *first four or five days* of life are after-effects of the management and health of the breeders or the management and hygiene of the hatchery. In the first section we will consider the commonest ailments of the chick.

Yolk-sac infection (also called 'mushy-chick' disease, omphalitis or navel infection

This is probably the commonest of all diseases causing early mortality. The signs are clear – the chicks so affected are usually normal to start with but make no headway and die mostly from about 24–72 hours after placement in the house. Examination of the chick shows a 'blown-up' abdomen, a wet or scabby navel and a very offensive odour which is virtually diagnostic. If the abdomen is opened it will be seen that the yolk sace is not absorbed but is filled with discoloured fluid, often brown, and infection may have spread throughout the abdominal cavity. The infective agents that cause the disease are bacteria and the ubiquitous *Escherichia coli* is frequently the main one involved.

Why does the disease occur? It may be due to a very substantial number of reasons ranging from mismanagement of the breeder flocks or bad hygiene, to poor hatchery standards or accidents in the control of the incubator or hatchery.

It is significant that we tend to find a much higher incidence in chicks hatched towards the end of winter or in the early spring when the health and condition of the breeding flocks are usually at their lowest ebb.

No treatment will save those chicks that are affected but it can be most useful to treat affected flocks since it may help to limit the spread of infection. The infected birds must release into the environment enormous numbers of potentially pathogenic organisms so the following measures may be rationally applied:

1. Increase the warmth in the house to assist the birds' resistance to infection and help to prevent huddling.
2. Have tests carried out in the laboratory to establish which organisms are involved and which treatment would be most appropriate.
3. Medicate the drinking water with a vitamin and antibiotic mixture.
4. Cull all abnormal chicks vigorously.
5. Make careful investigation of the hatchery or breeding organisation as to the likely cause. It is of great importance to do this and trace the source of the disease since if the breeding management is not informed of the trouble it may have no awareness of the problem and obviously no measures will be taken to stem its incidence.

Aspergillosis (or brooder pneumonia)

This is the most important but by no means the only disease of poultry caused by a fungus and leads to early death in chicks. The fungus, usually *Aspergillus fumigatus*, is one which occurs widely on farms and in the countryside and can therefore quite easily contaminate hatching eggs. The fungus in natural conditions needs mouldy, damp conditions in which the moisture content is more than 20 per cent – so it can be common, for example, if breeders are kept on damp litter or there is damp and dirty material in the nests. Hatching eggs can pick up the fungus in the nests or on the litter itself quite readily and the pathogen will then be taken through to the hatchery. If infection enters the hatchery and the fungus is not destroyed by fumigation, an almost explosive incidence can occur in the chicks and mortalities in the first few days of life may vary from a few up to 50 per cent or even more on occasions. Alternatively the chicks can themselves pick up infection from damp and mouldy litter used as their bedding.

The signs of infection are of pneumonia, the chicks gasping and struggling for breath with their mouth open. Post-mortem findings are of tiny yellowish-green nodules with a green furry extrusion in the lungs, bronchi, trachea and viscera, whilst the air sacs may be covered with thick yellow exudate. A laboratory will identify the fungus definitely – a sure diagnosis cannot be made by the naked eye or by the symptoms which may be similar to other diseases.

There is no specific treatment. If outbreaks occur in chicks they act as a warning of considerable importance that there are serious errors in hygiene and, or, management and it is vital that the areas of infection are located so that remedial steps can be taken. Control measures will include a culling of affected chicks, a more thorough disinfection of housing and equipment, removal of unsuitable litter or nesting material, and better cleaning and fumigation of hatching eggs and equipment and the hatchery. Obviously the precise steps to be taken will depend on the place where the infection takes place.

Salmonellosis

The group of bacteria known as *Salmonella* exists world-wide and is one of the greatest causes of diseases in poultry. In many countries, including the U.K., due to prolonged control measures, there are now few losses in poultry due to salmonellosis. Furthermore, it must be stressed that overriding emphasis is now attached to an attempt – which may well be achieved before long – to *eradicate* salmonella disease altogether from British birds because certain members of this species of bacteria are a principal cause of food poisoning in man. It will be appreciated that under intensive conditions there is an extra risk of contagion taking hold of many birds at one time so it is highly desirable that all possible sources of infection are removed. A few areas of the world – notably Scandinavia

– have been able to virtually rid themselves of salmonella infections in animals.

Types of salmonella infection

There are two forms of the salmonella species which are specific to poultry – these are *Salmonella pullorum* and *Salmonella gallinarum*. The former, perhaps the oldest of all diseases diagnosed in poultry, used to be called bacillary white diarrhoea (BWD), whilst *S. gallinarum* is commonly called fowl typhoid. In addition we have those diseases in poultry caused by other salmonella and coming under the general title of salmonellosis or sometimes termed avian paratyphoid. There are many different forms – at least 200 types of salmonella have been found in poultry – but the most important is *S. typhimurium*, which can be a virulent killer of chicks in particular.

Symptoms

The symptoms of the various forms of salmonella infection vary considerably. Mortalities are extremely variable, ranging from no more than one or two per cent up to as much as fifty per cent, or even more. If chicks, for example, are infected with *S. pullorum* disease via the infected dam and thereby spread through the hatchery, signs of illness will occur within a matter of hours of hatching. In cases where the infection is light and the environment and husbandry good, the losses can be modest but early losses from salmonella infections more usually run at between 10 and 20 per cent and extend over perhaps two weeks. Those chicks dying early on in life usually die quickly with few symptoms but as the disease proceeds, symptoms progressively appear of huddling, inappetence, constant cheeping, a swollen abdomen, and white diarrhoea. There may also be chronic cases in which numbers of chicks are stunted, scour and experience lameness due to swollen joints. It may be emphasised that these symptoms are much the same for all forms of salmonella infection in chicks and growing birds and it is quite impossible from these symptoms either to distinguish which form of salmonella is present or indeed whether it is a salmonella disease at all.

In the adult, acute salmonella infection is typically that known as fowl typhoid, due to *Salmonella gallinarum*. Signs that would lead to suspicion of an attack would be when a number of birds scour with greenish-yellow, evil-smelling droppings, have a greatly increased thirst, show very pale (anaemic) head parts and there are also a number of sudden deaths. As mortality may go as high as 75 per cent with 50 per cent being common, this is a disease which is hardly likely to be missed!

Post-mortem signs in the chick are not diagnostic but are at least strongly suspicious. If the bird dies at the hyper- (very) acute stage the signs are few, apart from congestion of internal organs, especially liver and lungs. In later cases, where the infection is only acute or sub-acute, the liver and other internal organs show characteristic yellowy nodules and the caecal tubes often contain semi-solid yellow 'casts'. The yolk sac may likewise be filled with a pus-like 'sludge'. In the adult, the post-mortem signs in pullorum or gallinarum infections are similar but are especially associated with the ovary in which the ovules are grossly misshapen and the contents are also abnormal. Other internal organs may be infected, including the heart which shows inflammation of the membranes around it (pericarditis). These diagnostic symptoms in the adult are only rarely seen with the other forms of salmonella.

The control of salmonella infections

With pullorum and gallinarum infections, simple and rapid on-the-spot blood tests have been in use for many years and can detect carriers with great accuracy. These tests have been enormously successful since they enable immediate removal of the carrier birds from a flock. Thus the two diseases caused by *S. pullorum* and *S. gallinarum* have been controlled with great success in many countries including the United Kingdom. Under these circumstances only a few isolated cases turn up from time to time detected by blood tests and regular testing may only be necessary on the foundation breeding stock.

Unfortunately the position with regard to other types of salmonella is entirely different and somewhat confusing. There are some 200 different serotypes that have been isolated from poultry. A few can be quite devastating in their effect, such as *Salmonella typhimurium*. Others are mild or cause no symptoms at all. In every case infection may spread to other livestock or man. It is probable, however, that the main risk is not that the poultry will infect other forms of life, but that other livestock will infect poultry!

In general the symptoms of salmonella infection in chicks are similar to those seen in pullorum disease and indeed it is only possible to distinguish the cause by laboratory examination. Post-mortem lesions do not enable more than suspicions to be raised though there is always the likely involvement of the liver which is usually enlarged and congested.

Salmonella and the consumer

Salmonella has long been known as one of the most important causes of food poisoning in man. In recent years it has been recognised that

Salmonella enteritidis is the most important serotype reported from human cases of salmonella food poisoning and it was claimed there was a clear epidemiological association of these cases with eggs, egg products and poultry meat. The British Government therefore introduced in 1989 a programme of measures for the control of salmonella in eggs and poultry.

The measures are aimed at reducing infection throughout the complete production chain, from the supply of feedstuffs to the production of healthy breeding stock and laying flocks.

Under the Zoonoses Order of 1989 all isolations of salmonella from samples taken from an animal or bird, or from the carcase, products or surroundings of an animal or bird, or from any feedstuff, must be reported to a Ministry Veterinary Officer. The Order allows Veterinary Officers to enter premises and carry out an investigation. Powers are available to declare premises infected, to prohibit the movement of animals, poultry, carcases, products and feedstuffs in or out of the premises, except under a licence – also serving notices requiring the cleansing and disinfection of premises where salmonella is known to have been present.

The Poultry Laying Flocks (Testing and Registration, etc.) Order 1989 requires owners of flocks of not less than 25 birds which are kept for the production of eggs for human consumption (including birds being reared for this purpose) of flocks of less than 25 birds the eggs from which are sold for human consumption, to take samples from their flocks and have them tested for salmonella at an authorised laboratory in accordance with the schedule laid down in the Order. Owners of flocks comprising 100 or more laying hens must register with the Agriculture Department.

The Poultry Breeding Flocks and Hatcheries (Registration & Testing Order) 1989 requires the owners of layer and broiler breeding flocks of 25 or more birds, and hatcheries with an incubator capacity of 1000 eggs or more, to register with the Ministry of Agriculture. Owners of breeding flocks and/or hatcheries must also take samples from birds or their progeny and have them tested for salmonella in an authorised laboratory in accordance with the schedules laid down in the Order.

Operation of the programme in respect of laying and breeding flocks

When an isolation of *S. enteritidis* is reported from a laying flock, either a flock producing eggs for human consumption or a layer breeder flock, a notice is served on the owner of the premises declaring it to be an infected place and prohibiting the movement of eggs, birds, food, etc. off the site except under licence. An investigation is then carried out in which a sample of birds is taken from each house on the premises and subjected to a detailed bacteriological examination. Investigation may also be carried out when there is evidence linking a flock to an outbreak of human food poisoning.

If *S. enteritidis* is isolated from the sample of birds taken from a house, all the birds in that house are compulsorily slaughtered and the owner is compensated. Birds may be slaughtered at an approved slaughterhouse and the carcases heat-treated to ensure their freedom from salmonella infection before they are sold. Once the birds have been removed the house is cleaned and disinfected. The house cannot be re-stocked until samples taken from the house and its contents are shown to be free of salmonella. During 1990 the arrangement described above for dealing with *S. enteritidis* also applied to *S. typhimurium*.

Broiler breeder flocks

Government measures to tackle the problem of salmonella contamination in poultry meat have concentrated on advice to caterers and consumers on correct handling and cooking practices, advice to local authorities and Official Veterinary Surgeons on hygiene in poultry slaughterhouses and cutting plants, the monitoring of feedingstuffs, the phased slaughter of infected breeding flocks and the operation by producers of a code of practice for the control of salmonella in broilers.

On 12 June 1990 the programme of compulsory slaughter of infected broiler breeding flocks was modified so that after that date an infected flock could remain in production until there was evidence of vertical transmission of infection to progeny through weekly hatchery monitoring carried out under the Poultry Breeding Flocks and Hatcheries (Registration and Testing) Order 1989. Once infection in a flock is confirmed, all hatchery eggs produced by that flock must be hatched separately from other eggs produced by other flocks. So far, only flocks infected with *S. enteritidis* and *S. typhimurium* have been slaughtered.

Feedstuffs

The Processed Animal Protein Order of 1989 requires all processors of animal protein to register with MAFF and to ensure that samples of processed animal protein consigned from their premises on any day are submitted to an approved laboratory for testing for salmonella. If a salmonella is isolated from the samples, no processed animal protein may be removed from the premises for incorporation into animal or poultry feedstuffs unless certain conditions specified in the Order are complied with. In addition, authorised officers carry out quarterly inspections involving the collection and bacteriological examination of samples of 20 days' production from each premises per year.

There are also more rigorous licensing conditions for the importation of animal and fish protein which were introduced in 1989 under the Importation of Processed Animal Protein Order 1981 (as amended).

The effects of the regulations

The effects of these regulations have been profound and there seems no reason why the poultry flock of the country cannot be rid of those forms of salmonella which cause disease in man. The cost to the poultry industry has been severe. The foundation breeding flocks have in any case been kept free of these serotypes of salmonella for a number of years and, as virtually all birds come from this stock, that would seem to be the end of the question. There is no doubt that British poultry products are second to none in their safety to the consumer but yet there remains a sense of unease about the matter. Our poultry may be free of these and other zoonoses but they will always be vulnerable to the introduction and re-infection from other sources, such as wild-life, over which we have no control, the whole environment around them and above all *homo sapiens*. In Sweden they have had no virulent salmonella in their birds for some time but a higher proportion of the human population suffers from salmonella infection than in the U.K. Scandinavians tend to holiday in the warmer countries around the Mediterranean and elsewhere where salmonella is rife, and some become carriers. The same circumstances obviously apply to British holidaymakers. Also, animal products, including poultry, enter the U.K. from European Community countries in an almost unrestricted way, and several E.C. countries have no salmonella control programmes. Thus with people and animal products constantly putting our birds at risk, the U.K. Government's policy is at the least worrying.

It should never be forgotten that if food is properly prepared and handled the risk of food poisoning is almost completely removed. More emphasis in this direction, together with improved Codes of Practice on the agricultural side, would in my view have been adequate and would have saved the compulsory slaughter of many millions of chickens. Politicians, vested interests and the media have been responsible for generating an emotional rather than an objective reaction, and the cost has been unnecessarily high.

Respiratory diseases of chicks

With the increase in the intensity and size of poultry units has come such an upsurge in the incidence of respiratory diseases that these, as a group, without doubt form the most important reason for ill-health in poultry in almost all parts of the world. It is unusual nowadays to find respiratory disease in an uncomplicated form. The usual pattern is for one or more primary virus infections, acute or chronic, to be followed by secondary bacterial invaders in which strains of *E. coli* usually predominate. Preventive techniques are all important since once respiratory disease has developed in a flock, and especially the young bird, losses by mortality, or poor growth and food conversion can be so severe that it may be extremely difficult to stop a catastrophic economic loss.

The approach to preventive losses cannot be easily generalised because the types of infection and their incentive vary a great deal geographically. If there is a high incidence, say of Newcastle disease and/or infectious bronchitis, two of the main diseases, a central plank in the preventative policy will be to protect the birds by vaccination. The strains of vaccine to be used and their frequency of use will be dependent on the local challenges that predominate and the parental immunity of the chicks. This parental immunity may have been derived from infection in the parent flock or from a suitable vaccination policy. As well as protection by vaccination, which will always be concerned with protecting birds from the primary viral agents, there may also be useful ways of protecting the birds against the secondary bacterial infections by medication of the stock, either in the food or drinking water, with antibiotics or chemotherapeutic agents. And, finally, all these measures will be supported by good husbandry methods; for example, by applying warmth, improving air flow, adding litter and changing lighting systems. We are faced once again with the salient fact that in many cases of respiratory infection it will be management rather than anything else which will influence its effect on the bird.

Infectious bronchitis

Universally this is the most widely experienced respiratory disease and is caused by a highly contagious virus. There are a number of strains and whilst most affect primarily the respiratory system there are some varieties which now seem to be on the ascendancy in many areas of the world which also have a serious effect on the kidneys, causing what is known as the uraemic form of the disease.

Symptoms. The most noticeable sign in the young chick is a large amount of respiratory noise. Birds sneeze, gasp and shake their head; and in a large flock the best way of detecting early signs is to 'look and listen'.

Enter the house very quietly, or if the birds are disturbed by your entry, wait until they are still. If you look you will see their heads shaking and probably an uneven spread of birds. If you listen the 'snicking' of the birds will be clearly heard and is almost symptomatic. Early observational diagnosis is all important because it enables immediate use of any preventive medication to be instituted. If the disease is allowed to advance unchecked the respiratory noise will get worse, there will be eye-watering, nasal discharge and increasing mortality and secondary infections which may take over, completely masking the original infection.

If the only trouble is the infectious bronchitis virus, this could be over within a week to a fortnight, but it is rare for there to be such a happy outcome. Nevertheless, if the disease is uncomplicated the effect can be mild and recovery can be complete, rather like the common cold in man. What usually happens in practice is that as soon as the birds show signs of sickness they start huddling together and this provides the ideal media for the propagation of other infections; viral or bacterial, but often dominated by *E. coli.*

In older birds infectious bronchitis disease can have two profound effects on the female. Infection in the immature female can prevent development of the oviduct partially or completely so that the bird will lay few, if any, eggs. In the other case, if the flock is adult and laying it will lead to a massive drop in egg production, which may partially recover, but in addition there will tend to be, after a time, the production of poor quality eggs and egg shells. Eggs are often of strange shapes, the egg shells are ridged and lack pigment and the white of the egg is watery.

When layers are affected, egg production may fall to a few per cent and it *may* then recover to normal over a period of weeks. However, my own experience is that this seldom happens and this fact, taken together with the lowered egg quality, means that the loss to an egg producer can have very serious consequences. It has even worse consequences for the poultry breeder because a bronchitis infection during production can lower not only egg numbers and quality but will additionally adversely affect fertility and hatchability. All these points emphasise the importance of giving susceptible birds protection by a complete vaccination programme if there is a fair risk of a challenge with the field virus.

Post-mortem signs in chicks affected with the respiratory form will include the detection of catarrh in the trachea, congested lungs and cloudy air sacs. When the kidneys are affected, they appear considerably enlarged, and white with the deposition of urates. Sometimes urate deposits are seen throughout the body cavities. The effect on the oviduct is to reduce it in size, sometimes so much so that it is almost absent, whilst with the mature bird the mis-shapen ova are diagnostic.

Diagnosis. The symptoms of infectious bronchitis after an incubation period of usually 1–4 days appear to be so reasonably straightforward that they would allow a clinical diagnosis to be effected without laboratory tests, but because the occurrence of the disease is rarely without other intruders, it is often necessary to have samples dealt with at the laboratory to be certain. For example, as shall be seen in the ensuing sections, the milder forms of Newcastle disease or mycoplasmosis, can produce rather similar symptoms. In any case, once the secondaries have come up to have their damaging effects, the diagnostic signs have all been very well masked. Thus plenty of use is made of the various tests which are either by virus isolation or by specific antibodies in the sera. The great advantage of having a positive diagnosis done is that while it is most likely that the result will not be in time to influence the treatment to be used in the flock in question, it may be possible to plan or modify future control measures in the light of this information.

Control. Details of the approach to the control of the disease are given at the end of the section on Newcastle disease because it is best to consider the control of these two rather similar diseases together.

Newcastle disease (ND)

Just like infectious bronchitis, Newcastle disease tends to occur throughout the world wherever there are large poultry populations. Whilst the first outbreak that was reported and described occurred in Newcastle, England, in 1927, this came as an importation of the most virulent form of the disease which occurred in the Far East. The first outbreaks that occurred in those days were nearly all extremely virulent and killed off virtually all affected and in-contact birds. Since that time, various other forms of the disease have occurred, many being much milder in extent and others somewhere between the two extremes. At the time of writing the most virulent cases appear to occur in the warmer countries, and the milder cases in the more temperate climates. The last serious outbreak in the United Kingdom of really virulent ND started in Essex in 1970 but the so-called 'Essex 70' strain which could cause almost 100 per cent mortality has now disappeared from the British Isles and indeed no ND infection of any type is present. Any outbreak would be countered with the slaughter of all infected flocks and possible contacts that could be incubating the disease. Vaccination might be used under official direction in a ring around an infected area to prevent the infection escaping from a small focal point.

Symptoms. The incubation period of ND is usually 2–7 days but may be prolonged up to three weeks. Normally the first signs in a flock are an unusual lassitude and indifference to food and water throughout the flock. This complete lack of the normal reactions of lively birds is quite symptomatic. There may be signs of respiratory distress, some birds apparently struggling for breath, with weird high-pitched 'cries' issuing forth in isolated cases – a symptom which does not seem to occur with any other disease. There is likely also to be a green watery diarrhoea, and most important in suspecting the disease from the clinical signs is the appearance of nervous signs – which does not occur with the other respiratory diseases. The nervous signs show up as a drooping of either one or both wings, incoordination, falling over and paddling of the legs and twitching and twisting of the neck muscles.

If the three main groups of symptoms I have listed are found in a flock then the presence of ND would be fairly certain, but a diagnostic check should always be made and in any case it is unlikely that the pattern would be quite so logically presented as I may have implied. Only some of the symptoms might be seen. Confusion can occur with several other diseases and in addition if some protection has been afforded by parenteral immunity or partial vaccination the end picture would be a very different one. For example, in layers the only effect might be a drop in the egg production and egg quality without a subsequent regain to normality. It will already be apparent that confusion with infectious bronchitis could be very easy, and there are many more diseases which could be confused with it.

Post-mortem signs. Some or all of the following signs may be seen: the respiratory tract is usually inflamed, the air sacs are cloudy and there are small haemorrhages – known as 'petechiae' – on the heart, and proventriculus (especially). There is often a great deal of caseous (dried pus-like) material in the trachea and bronchi and haemorrhages can be seen on the mucous membranes of the respiratory system.

Diagnosis. Clinical symptoms are described, together with the most prominent post-mortem signs as have been indicated, may be sufficient to be almost sure but it is nearly always necessary to have laboratory tests done. This may be done by virus isolation and/or serological testing, mostly using the 'haemagglutination inhibition' test.

Control measures. If we start with the chick, in most cases where the disease is prevalent it will be necessary to vaccinate all parent flocks very thoroughly. It is then likely the chicks will be born with a good immunity. There may not seem, therefore, very much point in an immediate vaccination and whilst this is not entirely untrue there are good practical reasons in large flocks to give some protection almost as soon after hatching as possible. One reason is that in large houses of many thousands of birds there will almost certainly be some chicks which do not have an immunity. Thus an early vaccination will protect them. Furthermore, if an early vaccination is given, it has been found that later vaccinations will cause less stress. The problem of stress cannot be exaggerated. The vaccines now used on chicks by water or spray administration are very effective but they are live and mildly virulent. Properly used they will cause no harmful effects but if incorrectly or carelessly used they are quite capable of causing a severe reaction which may be followed by secondary infections that can be almost as bad as the disease.

Protection of the birds against Newcastle disease and infectious bronchitis
For Newcastle disease there are three widely used vaccines – two being live vaccines (Hitchner B_1 and La Sota) and one an inactivated oil-based vaccine. The first two are administered by spray, drinking water, or occasionally eye-drop, whilst the oil-based vaccine is given by injection. Hitchner B_1 is a safe 'gentle' vaccine which when given by mass techniques, especially a spray, provides instant protection but protects for only a few weeks. La Sota is a little more stressful but is still mild and will protect for up to 12 weeks in the maturer bird. Now there are additional 'clone' vaccines with extremely mild effect but better protection. The oil-based vaccine by injection is most frequently used in the older bird – commonly at point of lay – and will protect for a laying season but does not give immediate protection. Nevertheless it seems that within this 'armoury' of vaccines the protection needed for any set of circumstances can usually be found, though in countries where the most virulent forms of Newcastle disease are present, it is usual to have some of the more powerful vaccines available, such as the Komarov. These have the disadvantages that they cause very considerable stress on the birds – hence most countries will not license their use.

There are two commonly used live vaccines to protect birds against infectious bronchitis. These are the relatively mild Massachusetts type H-120, which is used early in life and gives immediate and limited protection, and the H-52, which gives a much longer protection and is used to cover the laying period. It must be emphasised that full protection must be provided at all stages during development, not only to ensure the birds throw off any challenge that might cause respiratory disease, but also because of the danger that if a potential layer or breeder contracts the disease during its development, the reproductive organs can be so damaged that egg production will be profoundly affected. Alternatively, an inactivated oil-based vaccine may be given to protect adult birds.

In table 11.1 are given certain programmes of protection which can be used either as general routine protection where indicated, or for special emergencies or where there are severe challenges. It is as well to emphasise that even in the presence of Newcastle disease it is possible to spray birds with a large droplet (about 50 μm size or above), and this may arrest the disease on the site since the vaccine particles have a 'blocking' effect by giving localised protection via the cells of the respiratory mucous membranes. In this connection I would also emphasise the virtues of the large particle size spray, such as the 'Turbair', as compared with the small particle aerosol which is best *not* used. The latter penetrates deeply into the respiratory system and can cause a serious reaction. A method of using infectious bronchitis vaccine at day-old – if desired in the hatchery – has been developed by using the large particle spray, with considerable success as it gives a minimum of reaction and a maximum of protection.

Mycoplasma infection (CRD)

Mycoplasma are a group of organisms that cause a great deal of disease in the animal population. They are not dissimilar to, but are still quite distinct from bacteria, being smaller than them but yet being free-living, which distinguishes them from viruses.

Most of all in livestock they are associated with respiratory diseases and the long-known poultry disease known as 'chronic respiratory disease' (CRD) is primarily caused by *Mycoplasma gallisepticum*. The usual signs of the disease in a flock are sneezing, snicking and coughing, inflamed and running eyes and swelling of the sinuses of the face. The lesions to be seen in the bird are likely to be confined to the respiratory tract and will consist of varying degrees of inflammation and exudates in the tract, lungs and the air sacs. The disease in itself can cause some mortality and loss of productivity but it is very often a precursor to the other respiratory diseases. It is more serious in its effects on the younger bird than the older one.

Table 11.1 Newcastle disease and infectious bronchitis vaccination by the use of a large particle spray, or via drinking water or by injection. La Sota vaccine is not currently allowed in UK.

Age	Method	Vaccine		Remarks
		Broilers	Breeders/layers	
1–4 days General sensitising dose	Spray	B_1 & H–120	B_1 & H–120	Helps to reduce stress from subsequent vaccination
Primary vaccination 12–14 days	Spray	B_1 & H–120	B_1	No necessity to use both if risk is small
Secondary vaccination 22–28 days	Spray	B_1 or La Sota	B_1 or La Sota	
35–42 days	Spray	B_1 or La Sota	B_1 or La Sota	Final vaccination for broilers; unnecessary if low risk
10 weeks	Spray La Sota H-52 in water		La Sota or inactivated and H-52	
Just before point of lay	Spray or injection		La Sota or inactivated	Inactivated oil-based vaccines recommended for this application. La Sota live vaccine will give a much shorter immunity and necessitates re-vaccination during lay.
Boosting during lay	La Sota needs re-vaccinating after 5 months; no re-vaccination with inactivated			

Note: (1) Spray vaccination is generally more effective than drinking water application provided it is correctly applied. In those flocks in which coli-septicaemia is a major problem, spray vaccination can increase this problem and on such premises drinking water application may be used as an alternative.
(2) Spray vaccination works in two ways: it induces a rapid rise in antibody and also gives a useful measure of protection on respiratory surfaces themselves, known as a 'local block'.

Control. Mycoplasmosis is a less contagious and less acutely virulent disease than the worst forms of Newcastle disease or infectious bronchitis. There is no vaccine to prevent it but nevertheless it may be controlled very effectively in several ways. The most satisfactory way is to build up an entire breeding enterprise that is free of disease by blood testing and elimination of infected birds. There are good serological tests and this procedure has been so successful that most breeding stock in the U.K. and U.S.A. and western Europe are now free of *M. gallisepticum* infection. There are also several antibiotics that will successfully control the disease so these may be used in conjunction with a blood testing programme. Because the organisms can be transmitted from hen to chick via the egg, it is also recommended, in infected breeder flocks to treat the hatching eggs by dipping them in suitable solutions containing anti-mycoplasmal drugs.

A related mycoplasma, *M. synoviae*, is also a widespread cause of disease. Whilst it is primarily a cause of lameness due to the inflammation it causes in leg joints, it also retards growth and causes respiratory infections. Control measures are similar to those for *M. gallisepticum*.

Infectious laryngotracheitis (ILT)

Infectious laryngotracheitis (or ILT as it is more popularly termed for obvious reasons!) is a highly virulent viral disease which is world-wide in incidence but is confined to certain areas. Symptoms are usually of an extremely serious respiratory involvement with great distress exhibited by the birds as they struggle for breath, accompanied by coughing and rattling, and often with the bird depositing bloody mucous. There may also be less acute forms of the disease which are less severe but otherwise similar. Post-mortem examination of birds with ILT shows severe inflammation of the larynx and trachea with boody, cheesy 'clots' virtually filling the trachea. The virus that causes ILT does not affect any other organs and the birds die of asphyxiation.

Diagnosis can be made with a considerable degree of certainty from the symptoms, post-mortem picture and history and because the signs are not always as clear as has been suggested, and there may well be other complicating factors, it is advisable to have a laboratory 'check' on specimens and if necessary seek virus isolation or serological tests.

Treatment and prevention. There is no drug available that will affect the virus of ILT and the only effect of treatment with antibiotics and vitamin supplement will be to assist the birds' resistance to secondary infection. There are, however, good vaccines available to prevent the disease, the modern ones being the modified live virus which may be given by the eye drop route.

It is important to emphasise that 'recovered' birds may remain as carriers of the infection for the rest of their lives. Thus in an 'elimination' programme it is best to start by depopulating the entire site and then follow through with a most rigorous cleaning and disinfection programme. This is also one case at least where it would be advisable to leave the premises empty for a few weeks after the disinfection programme has been completed.

Infectious coryza (roup)

A bacterium *Haemophilus gallinarum* can cause a disease in poultry which is rather akin to the common cold in man. The symptoms are acute, lead to general inflammation, swelling and discharges around and from the eyes and the nose. It does not appear to be a very serious problem nowadays in intensive units but vaccines are made which are effective in preventing it. Treatment of the condition is to use antibiotics which the organism is known to be sensitive to after suitable test.

Gumboro disease (infectious bursal disease)

This is an extremely important virus infection in most areas of the world which was first reported in 1962 in the U.S.A. and very soon afterwards appeared in the U.K. and western Europe. The symptoms of the disease are rather indefinite but generally first show as an acute infection of young chicks – mostly broilers – of sudden onset, then a quick peak of mortality – followed by return to normality which, however, leaves a substantial number of birds stunted in their growth and very unproductive. Affected chicks show anorexia, depression, huddling, ruffled feathers, vent pecking, diarrhoea, trembling and incoordination. The most significant post-mortem sign is an extreme inflammation of the bursa of Fabricius (see Fig. 1.1) which will be greatly enlarged and oedematous. Haemorrhages may be seen here and in the muscles of the body, and the kidneys may be swollen, pale and containing urate deposits.

While the disease may not cause a high mortality – usually in the region of 5 per cent – the other effects of the disease must not be underestimated. The stunting in some of the survivors has already been stressed, but even more serious is the fact that the virus, by adversely affecting the bursa of Fabricius, causes profound damage to the whole immunological system of the chicken. The chick may suffer in consequence from other diseases, especially as it may now be unable to get proper protection from vaccines.

In many ways it is this last effect which is the most serious one and will be considered further in this chapter.

In the late 1980s a more virulent form of Gumboro disease became prevalent in various parts of the world and has now spread to virtually all of the larger poultry growing areas. The worst effects have been in broilers, and especially the cockerels, and mortality may reach 40 per cent when the birds are unprotected. Dealing with Gumboro disease involves three procedures. The first is hygiene to the highest standard, to minimise or at least delay the likely challenge on succeeding crops. The virus causing Gumboro disease is an extremely persistent one and it is unlikely it will be completely eliminated by cleaning and disinfection. Second, protection is given to the chicks by the use of a live vaccine administered by aerosol spray or in the drinking water. A wide range of vaccines of varying strength is available; the choice will be dependent on the degree of the local challenge expected, but at the worst it may be necessary to give the vaccine to a broiler flock three or even four times. Third, there are also inactivated vaccines for use in breeders so that they pass on a strong passive immunity to the chicks that should protect them for up to four weeks of age. Unfortunately the more virulent form of the virus breaks through the maternal immunity much earlier than this so the chicks may have to be protected from day-old. At the present time the disease is not being satisfactorily controlled.

The infectious stunting syndrome (malabsorption syndrome or runting and stunting syndrome)

A new condition appeared in several countries of the world several years ago affecting chiefly broilers and known as the infectious stunting or malabsorption syndrome. This disease is characterised by severe stunting of chicks arising very soon after they go into the broiler house – it is usually first noticeable at about ten days of age. Sometimes the condition is called 'helicopter disease' because feathering is very bad and certain of the wing feathers protrude horizontally from the bird. The syndrome may affect anything from 5 per cent to 40 per cent of the flock and by the killing age of about 50 days the affected birds will weight about half their normal weight. Deaths are relatively few; a characteristic post mortem sign in affected birds is a grossly atrophied pancreas, and it is assumed the agent causing this infection has a specific effect on this organ so that the digestion and absorption of food is badly affected. Thus, afflicted birds grow very slowly with an extremely poor food conversion efficiency.

It is known to be a highly contagious infection, almost certainly viral in origin but the causal agent or agents have not been positively identified.

There is no useful treatment for the affected birds – the sooner they are culled the better – but in flocks under risk the day-old chicks may be given preventive broad spectrum antibiotic treatment and soluble vitamin supplementation. To eliminate infection from the site apply the most rigorous cleaning and disinfection programme and in my experience even in earth-floored houses the disease can be eliminated. It is another example of the beneficial effect· of a depopulation and disinfection programme between batches.

In many cases it seems that broiler groups which practice good hygiene stay almost clear of the disease, though side-by-side in the same area, broiler companies with bad hygiene practices suffer badly.

Clostridial diseases

There are a significant number of clostridial bacteria which are capable of causing disease on the poultry farm. Clostridial bacteria are a large group which are capable of 'building up' as a problem, if the site is kept unhygienically, is overstocked or is badly constructed. Diseases caused by clostridia are malignant oedema (or gangrenous dermatitis), necrotic enteritis, and botulism. Clostridial organisms may also be found in association with many other diseases as 'secondaries', for example in yolk sac infection. There is also strong evidence that the weakening of the birds' immunity by Gumboro disease may be the forerunner to certain of these, especially malignant oedema.

Malignant oedema

Malignant oedema (or gangrenous dermatitis) is an especially objectionable and serious disease of broilers, caused by *Clostridium septicum*, which has become much more common. It occurs towards the end of the growing cycle and may cause a massive mortality – even up to 50 per cent – of the birds. The effect is to cause gangrene of the muscle and of the skin so that the birds in a sense 'fall apart' whilst still alive.

It is often stated and believed that malignant oedema is associated with bad hygiene practices, the re-use of old litter for example, and also overstocking and bad environmental conditions. There is probably a measure of truth in this but the disease can certainly occur without any of these being obvious and I have seen its devastating effects in flocks which are kept impeccably. What is known, however, is that it often happens after birds have had their immunity suppressed with diseases, such as Gumboro disease, or where other debilitating diseases such as inclusion body hepatitis (see next page) have occurred. Also it should be noted that while *Clostridium septicum* may be a main infective agent with malignant oedema it is by no means the only one involved and *Staphylococcus aureus* is often the predominating bacteria. Thus whilst penicillin-G is the antibiotic of choice to deal with this infection where clostridial organisms are involved, it may be less effective against staphylococcus. It is also important where outbreaks occur to consider carefully the programme for controlling Gumboro disease because outbreaks of this condition so often precede the attacks of malignant oedema. Perhaps the most effective preventive medication is with the synthetic penicillin 'Penbritin' (SmithKline Beecham), as this has a wide spectrum of activity including both clostridial and staphylococcal species of bacteria.

Necrotic enteritis

This is also a disease which is more usually found in broilers than in other types of chicken though not exclusively so. The clostridium involved is *Cl. welchii* and the condition may lead to a mortality of about 5 to 10 per cent. Lesions are confined to the small intestine which is grossly inflamed and distended.

This clostridial disease, like others, has often been found in poultry houses with earth or chalk floors where the disease organisms can 'build up' in certain areas which are probably inadequately cleaned or disinfected. Fortunately earth floors are becoming more of a rarity but where they are used it is worth emphasising that they can still be disinfected reasonably adequately with the modern materials available (see chapter 8). Preventive or curative treatment is effective using appropriate penicillin preparations.

Botulism

This is another disease of poultry caused by a clostridium (*Cl. botulinum*) which has occurred predominantly in broilers and can cause heavy mortality. I have seen cases of 20 per cent losses in birds near the killing age of 45 days. The course of this disease is rather different from other clostridia insofar as the organisms multiply in the gut releasing a toxin which causes paralysis of the muscles especially of the head and neck.

Fortunately it is relatively uncommon. The type of clostridium causing botulism in chickens is *not* the same as that which causes the disease in man. There is no treatment for poultry suffering from this condition but the only cases I have seen have been associated with dirty sites which are badly managed.

Inclusion Body Hepatitis (IBH)

IBH is a relative newcomer to the poultry field, also associated mainly with broiler chickens. It occurs generally in the age range 30–50 days. It has no special symptoms except that in a flock which is probably growing very well and apparently in the best of health, a sudden heavy mortality takes place which may give a loss of up to about 10 per cent. Post-mortem signs are confined usually to the liver and kidneys, the former being much enlarged and containing many haemorrhages, whilst the kidneys are also enlarged with haemorrhages throughout. A laboratory examination should be made to confirm the presence of this disease which is apparently caused by a virus.

Fatty liver kidney syndrome (FLKS)

This is an interesting and nowadays fortunately relatively uncommon disease of poultry of about 7–35 days old. The symptoms are indefinite – usually there is a rather higher mortality than there should be, and a few birds will be seen lying face downwards and apparently paralysed. Post-mortem examination of the affected birds shows a pale, swollen liver with haemorrhages, the heart also pale and the kidneys swollen and usually inflamed. It should be noted that it is usually the best birds which are affected by this condition and it is associated with excessive levels of fat deposition.

Cause. It has now been established that FLKS is a disease induced by feeding high levels of wheat with lower than normal levels of protein so that the ratio between carbohydrate and protein is too wide (see chapter 4). It has also been confirmed that if diets containing these incorrect levels are fed, then supplementation of the diet to give levels of biotin as detailed in chapter 3, will prevent FLKS. Now that this fact is known, the condition is a rarity but it does occur from time to time due to omission by the compounders of added biotin.

Coccidiosis

Coccidiosis is undoubtedly the most important parasitic disease of poultry and yet, because of the successful development of drugs to prevent and treat the condition, it is no longer a disease to be feared. Nevertheless, because it is always a potential threat which can cause serious loss, it is essential that the poultryman knows when to suspect it and what to do in an emergency. He should also be aware of the range of drugs available and how they may be effectively used.

Coccidosis of poultry is caused by protozoa (single-celled parasites) which live in the lining (epithelium) of the intestine. Those affecting poultry are of the species Eimeria. The life cycle of the various species in poultry is as follows.

Table 11.2 Coccidiosis in poultry

Species	Age group most often affected*	Key symptoms	Post-mortem signs
E. tenella	4–8 weeks; cases reported as early as 7–14 days.	Bloody droppings; marked drop in feed consumption; emaciation; high rate of mortality; lethargy very noticeable.	Caeca filled with blood; cheesy caecal core streaked with blood.
E. necatrix	6 weeks to 4 months; cases as early as 2 weeks; variable mortality.	Feed consumption drops; birds droopy; loss of weight; drop in egg production.	Middle of intestine involved (longer section in severe infections); small, white spots interspersed with bright or dull-red spots of variable size on unopened gut wall; wall thickened (sausage-like swelling up to two times normal size); haemorrhage into intestines; caecal contents may be tinged with blood.
E. acervulina		Loss of appetite; drop in egg production; loss of weight; some diarrhoea; chronic condition in layers.	Upper half of intestine affected; numerous, grey-white streaks on gut wall visible.
E. praevox		Some diarrhoea; relatively harmless; no deaths.	Upper third of small intestine; some degree of inflammation.
E. mitis		Diarrhoea principal symptom; no mortality.	Entire small intestine may be affected; slight degree of inflammation.

E. brunetti	May affect young or old birds.	Diarrhoea; emaciation; some mortality in serious outbreaks.	Lower half of intestine affected; rectum, caeca and cloaca inflamed and thickened; entire mucosa may slough in severe outbreak.
E. maxima		Diarrhoea; droppings may be flecked with blood; loss of appetite; emaciation; some death.	Dilation and thickening of lower half of intestine; gut filled with pink mucus or greyish-brown exudate.
E. hagani	Older birds usually affected.	Some diarrhoea; relatively harmless; no deaths.	Upper half of small intestine; small, round haemorrhagic spots visible through serosa in duodenum.
E. mivati	May affect young or old birds.	Unthriftiness; severe infections cause marked morbidity; drop in egg production.	Early stages in anterior 1/3 of small intestine. Later in lower small intestine, caeca and rectum. Rounded lesions congestion and white opacities, blood-tinged to watery faeces.

* Birds of any age may be affected by any species

There is a massive multiplication of the coccidia (known as schizogony) and sexual multiplication (known as gametogony) within the bird; then a stage of development takes place outside the host in which the eggs that are shed in sexual development reach a stage when they can infect other birds, though they do not multiply. These are known as oocysts.

There are nine strains of Eimeria species which can cause diseases of the intestines in poultry. Each species causes rather different symptoms due to the Eimeria parasites of each strain having different favoured sites in the intestine. For example, the most common ones are *E. tenella*, which affects only the caecal tubes and *E. necatrix*, which affects mainly the middle and lower intestine.

Symptoms. The over-all symptoms may be one or more of the following: bloody droppings, high mortality, general droopiness, emaciation, a marked drop in feed consumption, diarrhoea and egg production drop in layers.

Table 11.2 summarises the differences between the main species. These details show that there are certain differences in symptoms but a skilled diagnosis is often required for any certainty and can be made from a detailed examination of the post-mortem material in a veterinary laboratory. It is very desirable that good use is made of these skills since the type of coccidia that is found will give guidance on the reason for the infection, how to prevent it and how to treat it.

Outbreaks of coccidiosis which occur may be treated with sulphonamides or combinations of sulphonamides and pyrimidines or amprolium, all given in the drinking water. Sulphamezathine or sulphaquinoxaline are usually given at 0.1 per cent and 0.04 per cent respectively in the drinking water for 5–7 days.

A wide range of coccidiostats is used to prevent the disease during the growing period. A selection of these is set out in Table 11.3. Changes of drugs are appropriate by careful planning to prevent resistence developing. Certain coccidiostats too must be withdrawn a specified number of days before birds are slaughtered to ensure there are no residues in the carcase. Because of this it may be necessary to use two anti-coccidial agents, even in the short life of a broiler, as it is risky to try and dispense with any prevention during the last 5 to 7 days of growth. Replacement birds should not be given an agent that completely destroys the parasites since no immunity develops. It may be surmised, correctly too, that the choice, dose, and programme for the use of these highly important agents demands a skilled knowledge both of the drugs and the particular circumstances of their use. A recent addition to the protective range has been a vaccine which is currently used for replacement birds.

Coccidial immunity

It is especially important with the disease of coccidiosis to know how immunity develops. If birds are to achieve immunity they must ingest a sufficient number of oocysts to produce some infection. These oocysts then go through several life cycles and multiply. Following exposure with a sufficient number of oocysts the bird will then become resistant to the disease. Immunity must be established for each individual species of coccidia because there is no cross immunity and thus birds can become immune to one species of coccidia yet still be vulnerable to others. Also, it is noteworthy that immunity is not permanent. It can be lost if the environment of the bird does not provide a sufficient level of re-exposure. In certain tests, loss of immunity waas demonstrated within as short a period as 10 weeks.

Two principles are used to achieve the gradual production of immunity as a means of controlling coccidiosis. The first is by the use of drugs at low level to reduce the severity of natural exposure and the second deliberate seeding of the litter with coccidia to provide a controlled exposure. It is much safer and more reliable to use anticoccidial drugs. With both of these methods it is possible to have an inadequate exposure with a resultant deficiency in immunity, or an excessive exposure leading to losses from the disease.

The most common technique to control coccidiosis is to give continuous preventive medication for a defined period with one of the anticoccidial drugs in the food. All broiler chickens are given this treatment and so are most replacement breeding and egg production stock, but adult birds are not usually given preventive medication, since they should have achieved immunity by then.

Withdrawal of coccidiostats

In practical terms the difficulty must be stressed of dealing with the use and withdrawal of coccidiostats in young rearing birds. The chicks will usually be put on a coccidiostat from hatching and this is likely to both control the disease and allow the chicks to build up an immunity so that by the time the drug is taken off – say, between 12 and 16 weeks of age – a satisfactory level of immunity has developed and the birds can safely withstand any subsequent challenge. Birds must be very diligently watched after the withdrawal period because they may suffer from an outbreak of the clinical disease if the challenge is too great, and will need treatment with one of the drugs or drug combinations mentioned above. The likelihood of a severe challenge may be related to the efficiency of management; for example, damp, wet litter following a period of satisfactory dry conditions might favour a sudden outburst of coccidial activity.

Factors affecting infectivity

An initial factor affecting the course of the disease is the viability of the oocysts and their survival on the ground. Infected birds excrete enormous numbers of oocysts in their droppings in the fresh (unsporulated) stage. The oocysts are not capable of infecting birds until a further stage of development has taken place. If conditions are just right for sporulation it will take place within two days. Such conditions are warmth (25–33°C) with a high moisture content. However, if the environment is unfavourable, sporulation may take a long time or not take place at all: for example, at temperatures below 10°C, or very dry conditions. If the temperature is very high (45–50°C for a day or 56°C for an hour) the oocysts will be destroyed.

Once the oocysts have sporulated they are quite resistant to low temperatures (but not a hard frost) and are relatively resistant to dry conditions and many forms of disinfectants that would kill viruses or bacteria. They are, however, killed by temperatures above 56°C and by ammonia and methyl bromide gases.

In practical terms this means that oocysts can even survive outside the bird on the ground during winter and indeed they have been known to survive for up to two years. However, on deep litter the combination of dryness, ammonia production and bacterial decomposition results in the destruction of oocysts within about two weeks of being voided. Information on the appropriate use of disinfectants is given in chapter 8.

Blackhead (histomoniasis)

Blackhead is another disease caused by a protozoan parasite called *Histomonas meleagridis*. Whilst it is primarily a disease of turkeys it may also affect young chickens in a mild form. The symptoms include watery, yellowish diarrhoea, lethargy, weakness, loss of appetite. Post-mortem examination shows a marked inflammation, usually with cheesy cores, in the lumen of the caeca; on the liver a number of circular yellowish-green depressions or ulcers about 12 mm in diameter may be observed.

There are a number of excellent drugs to prevent or cure blackhead, including for example, Emtryl (Rhône Mérieux), Salfuride (Salsbury Laboratories) and Unizole (United Chemicals). The occurrence of blackhead in chickens indicates a need to improve hygiene and eliminate intestinal worm since their eggs may carry the blackhead parasites.

Viral arthritis (tenosynovitis)

A widespread form of viral arthritis is quite common in broiler chickens from about 4 weeks onwards. The signs are of serious lameness in the birds coupled with poor growth rates and bad food conversion efficiencies. In the affected birds the flexor tendon sheaths running along the posterior of the tarsometatarsal bones are seen to be swollen, giving the shanks a very enlarged appearance. The sheaths of the gastrocnemius tendons just above the hocks are also swollen. In advanced cases the tendons themselves may frequently rupture. At a post-mortem examination the sheaths can be seen to be oedematous, looking almost like gelatine in appearance. There is usually no pus in the inflamed tissues and the remainder of the bird appears completely normal.

It is often possible to make a diagnosis on the basis of the clinical signs but confirmation depends upon virus isolation and identification.

There seems little doubt that the virus causing this condition is widely spread in the world's poultry industries but only certain broiler flocks are without maternal immunity and are challenged. Thus the practical incidence of the disease is quite limited.

There is no cure for the condition but when flocks are affected the administration of an antibiotic with a wide spectrum of activity plus soluble vitamin preparations is a justified veterinary procedure in order to limit the effect of the disease, guard against secondaries, such as staphylococci and clostridia, and help to overcome the effects of inappetence and poor growth.

Epidemic tremor (infectious avian encephalomyelitis: IAE)

The condition of epidemic tremor is caused by a specific virus of chickens leading to nervous signs and symptoms in young birds. While birds of all ages can be infected it is only in birds up to about 8 weeks of age that the characteristic nervous signs are normally seen. In laying birds an attack causes a drop in egg production of about 10 per cent which is over soon, but a drop in hatchability also occurs.

The signs that would lead to suspicion of this disease are confined to the nervous system. Chicks show incoordinate movements, falling, collapsing, then becoming prostrate and finally dying. If the chick is picked up and rested in the palm of the hand the tremors of the muscle may be felt, which are quite characteristic of the disease. Usually about 10–15 per cent of the flock are affected and most of them die, though a few make a complete recovery.

A definite diagnosis is made on microscopic examination of the brain and central nervous system where specific changes take place.

Route of infection. The virus infects chicks in two ways. The first route is from the dam to the chick via the ovaries. After hatching the virus can enter by the mouth or respiratory system. If chicks are infected via the ovary, they contract the disease within about 7 days of hatching. If the chicks are infected by the oral route, they will become diseased from about 3 weeks onwards.

Control. Fortunately excellent live vaccines are now available to control epidemic tremor. The procedure undertaken is to vaccinate breeders to protect them against falls in egg production and hatchability and also to prevent the transfer of infection to their chicks. Vaccination is given between 10 and 16 weeks of age. Provided it is done within this period, protection should be satisfactory for the rest of the life.

Mycotoxicosis

It is worth noting that chicks of all ages may be affected with forms of the disease known as mycotoxicosis. This is really a group of diseases which arise from toxins released by moulds found in certain feedstuffs or in the litter of the poultry houses. Symptoms are usually indefinite but there may be a number of birds off-colour, obviously sick, droopy, eating and drinking little or nothing and several dying. Post-mortem signs are mostly confined to the liver which may be, in acute cases, congested, or in chronic cases pale, or yellowish-brown in colour and hard. Confirmation of the involvement of a mycotoxin requires detailed microscopic examination of the liver and of the feed the birds have been receiving.

There is no specific treatment but the administration of vitamins, especially B vitamins in the water, is warranted in an attempt to reduce some of the effects of the toxins. A number of chemicals are currently available for the treatment of either the food or the litter and housing, e.g. Kemin.

Marek's disease and lymphoid leucosis

Both Marek's disease and lymphoid leucosis are very important conditions of the chick which cause neoplasia (tumours) by two groups of viruses. Marek's disease is caused by a strongly cell-associated herpesvirus, whilst leucosis is caused by some of the so-called RNA tumour virus group of oncornaviruses. The major manifestations in both diseases are lymphoid neoplasia but in other respects the two conditions differ widely.

Marek's disease

Marek's disease was probably the most common of all poultry diseases before the relatively recent introduction of vaccines which has enabled its very successful control. It is a disease which causes proliferation of the lymphoid tissue, affecting most organs and tissues but with a special predilection for peripheral nerves. It affects chickens most commonly between 12 and 24 weeks of age, although it does occur occasionally in chickens from 6 weeks of age and sometimes, but rarely, in birds older than 24 weeks. The incubation period may vary from as little as 3 weeks to several months. There is one form of Marek's disease which is known as *acute Marek's*. This produces a very high mortality, up to 30 per cent is quite common, and on occasions even as much as 80 per cent. The disease may proceed in two ways: there may be a quick peak of mortality, rapidly falling away to almost nothing, all in a period of only a week or two or, alternatively, it may reach a peak quite quickly and then continue without much change for months; only a proportion of birds may show nervous signs of paralysis.

The alternative, and until recently the more common form, is that known as the 'classical' Marek's disease. In this, mortality is usually much less, rarely exceeding 10–15 per cent. In some cases mortality continues only for a few weeks and in others it may go on for many months. The symptoms differ widely depending on which nerves are affected. Most often there is a progressive paralysis of the wings and legs. Badly affected birds cannot stand and a characteristic situation is to find birds lying on their sides with one leg stretched forward and the other held behind. In other cases the neck may be twisted, or the respiratory system may be involved, or even the intestines.

On post-mortem examination, most typical signs with acute Marek's are diffuse lymphomatous enlargements of one or more organs or tissues – most often liver, gonads, spleen, kidneys, lungs, proventriculus and heart. Sometimes the lymphomas also arise in the skin associated with feather follicles and also in the skeletal muscles. In younger birds the liver enlargement is usually moderate in extent but in adults the greatly enlarged liver may appear identical to that in lymphoid leucosis.

Control. There is no treatment for Marek's disease. Control depends on the rearing of young stock in isolation from older birds, the use of genetically resistant stock wherever possible, and above all else, by vaccination. There are two main types of vaccine – either a modified field strain of Marek's disease virus, or a Turkey herpesvirus – which is a virus antigenically closely related to Marek's disease virus and which is present in most turkey flocks. It is not pathogenic for turkeys or chickens and can be used very successfully for vaccination.

There are two forms of the vaccines produced. In one form the vaccine is used in association with the cells, and in the other the vaccines are freeze dried. There is no evidence as to which is definitely the best and both can be recommended. The freeze dried vaccine which can be kept in an ordinary refrigerator is much easier to store and handle than the cell-associated form which must be kept in liquid nitrogen containers at −100°C. However, some poultry men consider the immunity derived from the cell-associated form is better, perhaps because rather greater care must be taken in its handling.

All vaccines are normally used once only at one day old, and this is most conveniently given at the hatchery. Within a week a reasonable level of protection is obtained and this lasts a lifetime. There have been some unfortunate 'breakdowns' in properly vaccinated flocks. There is no evidence that this is due to any fault in the existing advice or techniques but it is a warning that all the points I have stressed must be taken care of, viz. isolation of chicks, good hygiene at all stages, correct storage of vaccine, correct usage and administration of vaccine. Such has been the concern in the poultry industry about the 'breakdowns', that some organisations are now advising a second vaccination at about two weeks of age.

Leucosis

There are several different forms of avian leucosis. The commonest is lymphoid leucosis, in which there is usually an enlargement of the liver which is so grossly overgrown that it extends the full length of the abdominal cavity, whilst tumours may also affect any other organs of the body, the commonest sites being the spleen, kidney and bursa of Fabricius. Another type is erythroid leucosis which causes a serious anaemia due to the proliferation of excessive numbers of immature red blood cells. The liver, spleen and bone marrow become cherry red in this form. In contrast, in myeloid leucosis there is tumour formation in the liver and spleen, the liver often assuming the appearance of morocco leather and a granular texture. This is due to the presence of a large number of discrete and nodular tumours of a chalky or cheesy consistency.

The lesions are all caused by a group of closely related RNA tumour viruses. Virus strains produce predominantly one form of leucosis but many will in addition produce lymphoid leucosis. The viruses can be identified by the presence of a group specific antigen using the so-called Cofal test.

The course of the disease. Most commercial poultry have a certain degree of infection with some groups of the leucosis viruses. The viruses are excreted in the droppings and the saliva and can readily infect in-contact birds. In addition a proportion of the eggs are also infected to pass the virus through to the chick. Chicks receiving their infection through the eggs become tolerant to the virus and do not produce antibodies, whereas those infected after hatching respond with the production of viral antibodies. Chickens tolerant to the virus are more likely to succumb to disease, and if they do not they shed the virus in the eggs more consistently than chickens with a circulating antibody. Lymphoid leucosis is the main form of disease developing after infection with leucosis virus. It is stated, however, that only about 2 per cent of chickens die from lymphoid leucosis.

Control. There is no treatment, and because the infection is transmitted through the egg, management and hygiene methods are unable to control the disease. However, because the viruses are not highly contagious it is relatively easy to prevent infection of a flock from outside. If it is possible to free flocks from infection it is possible to maintain them and their progeny free of infection.

Genetic selection of stock resistant to infection with leucosis viruses creates a population resistant to infection with these viruses and also reduces the number of birds capable of supporting infection.

Fowl pox

Fowl pox is a virus infection of the pox group which causes lesions in the head and mouth of the bird. It is also possible to find lesions on the legs and mucous membranes such as the cloaca. The lesions on the head and comb are usually wart-like in nature. Those which occur in the mouth are diphtheritic and have the appearance of a cheesy membrane. The general symptoms depend on the area of the body affected. The birds will appear dull and when the mouth is affected may have difficulty in breathing and eating.

When the virus attacks the mucous membranes of the nasal and buccal cavities mortality can reach as high as 40–50 per cent. The virus causes a marked increase in the output of mucus and this is sometimes referred to as 'wet pox'. The less serious skin form of the disease does not lead to high mortality but can cause a reduced egg production in laying birds.

Diagnosis and treatment. The symptoms are certain enough usually to make a diagnosis of the disease, though if there are doubts a veterinary laboratory can confirm.

It is not easy to treat affected birds but it is possible to remove the diphtheritic membranes from the mouth and the area around and then treat with antiseptics and emollients. To prevent, it is best to use a live pox vaccine which can be administered either by wing web or feather follicle application. Fowl pox is a relatively uncommon condition in temperate climates under intensive conditions but remains as a common complaint in warm climates where the systems are less intensive and may not be scrupulously hygienic.

Egg drop syndrome 1976 (EDS'76)

Since the 1970s it has been known that a variety of viruses – so-called adenoviruses – can infect the chickens. Whilst they may have an effect at any time of life the most serious strains cause sudden and prolonged falls in egg production and a profound effect on egg and shell quality. One particular strain first caused problems in 1976 in broiler breeder parents and has since been designated EDS'76. The condition has also spread to other adult birds but it is in breeders that the effect has been most serious. The syndrome has apparently shown itself in a number of forms. There may be a failure of production to peak as it should, or there may be gradual or sudden drops in egg production. The egg production of an affected flock may or may not return to normal. There is also a tendency for lowered hatchability, poor egg shell quality and loss of colour to occur. Diagnosis of this condition requires expert examination since the symptoms can readily be confused with, for example, those of infectious bronchitis and Newcastle disease and perhaps other adenoviruses and infectious avian encephalomyelitis. A vaccine has now been produced which has been found to be capable of giving satisfactory protection, e.g. Nobi-Vac EDS'76 Vaccine (Intervet Laboratories).

Pullet disease ('blue comb'; 'monocytosis')

Pullet disease was once a very common condition of birds affecting them soon after they had come into egg production. It has all the signs of a viral infection but it has not been definitely established to be so. The symptoms that are seen are usually a marked fall in egg production together with inappetence and lethargy. Some birds may be found dead, dying too quickly to show symptoms, but in most, and especially later cases in the course of the disease, death is preceded by a short period of illness during which purple coloured combs and wattles are the most noticeable feature, together with diarrhoea and an extremely dejected appearance.

Post-mortem signs are especially associated with abnormalities in the kidneys, which are usually grossly enlarged with crystalline urate deposits. There are also degenerative changes of the birds' ovaries and a general congestion of the carcase, including a patchy discolouration of the pectoral muscles.

The only useful treatment is to administer tetracyclines in the drinking water, ensuring that this is freely available to all the birds. Restriction of food intake may be beneficial for a time and a 'traditional' remedy is to mix in with the diet 10–20 per cent of molasses and feed the whole diet wet.

Roundheart disease

This is a rather strange condition that occurs in layers on deep litter or straw between the ages of 6 and 12 months. Death is sudden and post-mortems show an enlarged, distorted (rounded) and discoloured heart. It is believed to be due to a toxic factor but so far none has been discovered and there is no treatment or preventative. Fortunately it is now uncommon.

Avian tuberculosis

Avian tuberculosis is a disease caused by a bacterium *Mycobacterium tuberculosis avium*. It is a contagious disease passed from bird to bird apparently by contact via the droppings of infected birds.

The disease normally affects older birds when there is a progressive loss of weight ('going light' as it used to be termed). The birds become unthrifty. The muscles of the breast are reduced in size exposing the sternum or breast bone. Appetite usually remains normal until the terminal stage of the disease. There may be some lameness and swelling of the joints. The comb and wattles appear pale in colour.

Post-mortem examination reveals lesions which appear as yellowish-white caseous nodules in the liver, spleen and intestines. Laboratory examination can confirm the presence of the disease.

No treatment is effective and the disease is best dealt with by good culling of the affected birds combined with a complete depopulation of the house and subsequent disinfection.

Fowl cholera (avian pasteurellosis)

Fowl cholera is caused by a bacterium *Pasteurella multicoida* and can be a very serious cause of disease in many parts of the world. Infection occurs by the respiratory or alimentary route which contaminates water, soil and feed. There is an incubation period of 4–10 days and the severity of the condition depends very much on the 'stress' of the birds by poor ventilation, overcrowding, and so on. Symptoms in the most acute cases include sudden death in large numbers with signs of cyanosis (purple or mauve colouring) and swelling of the comb and wattles. In the less acute stages there is swelling of the joints and legs causing lameness. Affected birds do not eat or drink and have difficulty in breathing. A thick nasal discharge will be seen together with a greenish-yellow diarrhoea.

Mortality may be very high indeed – even up to 90 per cent. Diagnosis is made on the basis of the symptoms and on isolation of the organism.

Treatment may be instituted by using antibiotics and drugs and prevention can be effected by vaccination.

Ascites, Oedema or 'Water-belly'

Ascites in broilers is an increasingly common condition. The symptoms are a blue comb, feathers ruffled and a grossly enlarged abdomen with fluid. After death if the bird is opened large quantities of brown-yellow liquid emerge.

Accumulations of fluid, or ascites, occur because the circulation of the blood system may in some way be impaired. It is believed this is primarily due to the very fast growth of broilers, together with insufficient ventilation and too high a stocking rate. Ascites is also associated with *E. coli* septicaemia, due no doubt to reduced cardiac efficiency. Sometimes losses can be very high and consideration should be given to the following measures to prevent losses:

- Reduce stocking density and increase ventilation.
- Feed a ration of reduced energy and protein to slow the growth rate slightly.
- When an attack occurs treat as for *E. coli* septicaemia with the same medicines.

Turkey Rhino-Tracheitis

This is a relatively new disease of turkeys which has a serious effect in the areas where turkeys predominate. In this condition the birds develop sudden respiratory symptoms with discharges from the eyes and nose and distensions of the sinuses. The most common age of infection is between about 8 and 16 weeks and usually nearly all the flock is affected. Whilst it is possible for flocks to recover quite quickly, infections such as *E. coli* may cause great havoc and losses can easily reach 50 per cent. In laying turkeys there is a severe loss in egg production, which recovers quite slowly.

The main cause of TRT is a pneumovirus which is immuno-suppressive and highly contagious. Unfortunately the disease is also able to infect chickens, both adult flocks and broilers. In adult flocks it causes a severe loss of egg production and in broilers 'swollen heads' with secondary effects.

Vaccines are under development and are obtainable for protection with variable success. Improved environmental conditions will assist, together with treatment for the secondaries in the usual way.

Duck Virus Hepatitis (DVH)

This is an acute, highly contagious disease of young ducklings aged from 2 days through to about 4 weeks. The effect is very sudden and the duckling usually dies within an hour or so of the onset of the symptoms – if they are seen at all. Losses can easily be as much as 90 per cent of a flock, especially if the birds are young. Post-mortem signs are in the liver, which is enlarged and haemorrhagic.

The only useful protection from DVH is a vaccine. The most satisfactory procedure is to vaccinate the parent stock, who will pass on protection to the progeny which should be sufficient.

Duck Virus Enteritis (DVE)

DVE or duck plague is a further virus infection which is very contagious and affects ducklings between 2 and 6 weeks of age. Symptoms are straightforward – heavy losses up to 100 per cent sometimes, diarrhoea, great thirst and haemorrhage throughout the organs of the body. Protection is by live strains of the vaccine.

Intestinal parasites

There are four main intestinal parasites of the chicken, as follows.

The tape worm (Davainea proglottina)

Affected birds lose weight and are listless; breathing is rapid and feathers are ruffled and dry. In laying birds there is reduced egg production. At post-mortem the intestinal mucosa may be thickened and haemorrhagic and the tape worms are seen. For development of the life cycle the intermediate hosts – such as snails and earthworms – have to be present. Specific treatment should be used where the worms are found. It is understandably only likely that infection will take place under extensive or semi-intensive conditions.

The large round worm (Ascaridia galli)

This is a large worm up to 80 mm in length which causes loss of condition, reduced egg production in layers and it can even cause death if infection is heavy. Retarded growth, listlessness and diarrhoea are the clearest symptoms. At post-mortem the large round worms can be seen in the middle part of the small intestine. The disease is uncommon under hygienic conditions but certainly occurs if the litter is wet and birds are overcrowded.

The hair worm (Capillaria)

The hair-worm is a small parasite which affects the upper part of the alimentary tract causing diarrhoea, weakness, reduced egg production and anaemia. Affected birds are listless, weak and lose weight. Hygienic measures and treatment will be as with the other worms.

The caecal worm (Heterakis gallinae)

This is a worm of about 10 mm size which develops in the caeca, affecting chickens and turkeys. Perhaps its most important 'function' is as a transmitter of the parasite causing blackhead (histomoniasis).

Treatments

There are adequate specific treatments to control the round and tape worms. For example, piperazine products are safe and effective against roundworms (see table 11.3 for more information on products).

External parasites

Poultry can be infested with a variety of external parasites such as lice, fleas, and various forms of mites, such as red mite, the depluming mite and scaly leg mite. These mites all burrow into various parts of the birds' skin and create considerable damage. The presence of all these conditions is rather uncommon under modern intensive management, and if general standards of hygiene are high it is doubtful if they will be seen at all. External parasites are more commonly associated with older birds and old-fashioned systems of continuous production. If they do occur, then it is advisable to deal with them energetically as they will have a highly debilitating effect on the birds and may also be associated with the spread of other diseases.

Treatment will depend on the administration of an assortment of effective products, which are listed in chapter 8. Of course it need hardly be emphasised that if birds are badly infested, especially with the mites, they should be culled as they are unlikely to respond very effectively to treatment, and they will remain as serious sources of infection.

Common poultry diseases and their treatment

The poultry industries of the world are served by an impressive array of companies producing advanced pharmaceuticals, hygiene products and biologicals. It is no exaggeration to say that without the development of these products modern poultry production as we know it could not exist. However, it is a vital corollary of this fact that the poultry farmer knows generally what is available and makes use of the correct product without delay. Successful enterprises are those which react quickly to any risk or appearance of disease, to prevent or treat. Products can be remarkably cost-effective when they are used in the right circumstances at the correct time. The wise poultry farmer will always consult his professional adviser when he has doubts on the diagnosis or treatment of a disease, or the preventive measures necessary to ensure good health. Table 11.3 lists the most common poultry diseases and gives information on the measures available for their prevention and treatment.

Examples are given of the available commercial products but the absence of some does not imply they are not just as good, merely that it was impossible to include everything worldwide. Most countries have a reliable system for testing and licensing vaccines and medicines and it is advisable that only officially approved and tested products are used.

There are many 'combined' vaccines for poultry incorporating two or three different vaccines and wherever these can be used they are preferable to using vaccines separately. Combined vaccines reduce stress and improve and accelerate the development of immunity.

Table 11.3 A chart of poultry diseases, their prevention and treatment

Parasites and Disease	Prevention	Treatment
Adenoviruses (including egg drop syndrome (EDS '76), inclusion body hepatitis (IBH) and turkey haemorrhagic enteritis (THE))	Vaccine will protect against one type, EDS '76 (egg drop syndrome 1976), i.e. Nobi-Vac EDS '76 Vaccine (Intervet Laboratories) and Binewvaxi Drop (Rhône Mérieux) combining egg drop syndrome, infectious bronchitis and Newcastle disease vaccines, or Dindoral SPF (Rhône Mérieux) for turkey haemorrhagic enteritis alone.	Non-specific measure such as vitamin, mineral and electrolyte mixtures and often broad spectrum antibiotics will help in the bird's recovery.
Ascites in broilers	Reduce stocking density. Increase ventilation. Improve litter provision. Lower diet intensity to slow growth.	Similar to *E.coli* septicaemia which often occurs at the same time. Potentiated sulphonamides especially useful, e.g. Uniprim 150 in feed (Cheminex Laboratories) or Duphatrim in feed or drinking water (Solvay Animal Health).
Aspergillosis (brooder pneumonia)	Avoid all damp litter and nest material. Check hygiene in all areas of the hatchery.	Treat the whole environment with a disinfectant known to be active against fungi, etc. e.g. Farm Fluid S (Antec International) 1:400 or Microsol (Micro-Biologicals) 1:100.

Avian clostridial diseases including necrotic enteritis, botulism and gangrenous dermatitis	Hygienic measures within and around buildings require to be improved and there is a need for an improved disinfection and fumigation programme. Disinfectants and fumigants should be used with their activity specified against clostridial bacterial spores which are resistant to destruction.	Soluble penicillin or semi-synthetic penicillin in drinking water at 0.5 g per 5 litres of water, e.g. Clamoxyl Soluble Powder (SmithKline Beecham) 20 mg active agent (amoxycillin per kg live weight) or Amoxinsol 50 (Univet) 15 mg amoxycillin per kg live weight.
Avian influenza (fowl plague)	Blood testing or virus isolation to be followed usually by slaughter of positive flocks, but vaccines can be used as alternative.	No useful treatment other than supportive antibiotic and vitamin adminstration.
Avian leukosis	Blood testing and elimination of infected stock with selection of genetically resistant strains. Also improve all measures of hygiene and isolation.	None.
Avian mycoplasmosis (chronic respiratory disease – CRD)	Normally eliminated from breeding flocks by blood testing and removal of reactors. Also by medication, e.g. with Linco-Spectin 100 (Upjohn). One pack incorporating 100 g of antibiotics per 200 litres of drinking water; or Tylan Soluble (Elanco Products) 2.5 g per 4.5 litres. Vaccination sometimes used but only partially successful.	Similar to preventive treatment. In addition may use tetracycline; macrolides, amino-glycosides and quinolones.
Avian pasteurellosis (fowl cholera)	Vaccination by injection, e.g. Pabac Pasteurella Vaccine (Solvay Animal Health) and Pasteurella and Erysipelas Vaccine (Hoechst).	Chlortetracycline, oxytetracycline, amoxycillin, e.g. Amoxinsol 50 (Univet) 15 mg per kg body weight or Clamoxyl Soluble Powder as recommended (SmithKline Beecham).

Parasites and Disease	Prevention	Treatment
Avian tuberculosis	Test for reactors and cull infected birds. Ensure hygiene is improved as organism causing disease is persistent.	Not advocated.
Coccidiosis	Medication of ration with a variety of cocciodiostats, e.g. Elancoban (Elanco Products) 100 g per tonne of feed or 500 g of 20% Premix, broilers only, or Salinomycin (Hoechst) 60 g per tonne of feed, broilers only, or Coyden 25 Premix (Rhône Mérieux) containing 25% Clopidol, or Deccox Decoquinate (Rhône Mérieux) 30–40 g per tonne, or Cycostat 66 (Cyanamid) 500 g of mixture per tonne of feed, or Dot Pure (Roussel Laboratories) up to 125 g per tonne of feed, or Baycox (Bayer UK) – short administration via drinking water. Also vaccine, e.g. Paracox (Pitman-Moore).	Amprol-Plus (Merck Sharp & Dohme) contains amprolium + ethopabate 400 ml in 100 litres of water for 5–7 days, then 100 ml per 100 litres for 14 days, or Microquinox (Micro-Biologicals) 300 ml to 200 litres of drinking water on 3–2–3 system of dosing.
Duck virus hepatitis	Strict isolation in first few weeks of life and vaccination of parent stock to provide parental immunity, e.g. Hepatovax Live (Rhône Mérieux).	Serum therapy possible and occasionally used.
Duck plague (duck virus enteritis)	Cleanliness and hygiene plus vaccination, e.g. use Duck Plague Vaccine Nobilis, or Vaxiduk (Rhône Mérieux).	None.

E.coli infections	*E.coli* is usually a secondary infection, thus main efforts must concentrate on prevention of primary disease with vaccination, therapy or good husbandry. Useful medicines are: furazolidone in feed at a level of 200 g per tonne of feed. Also potentiated sulphonamide, e.g. Uniprim (Cheminex) or Synutrim (Peter Hand), each at 1 kg per tonne of feed or a variety of water soluble preparations such as chlortetracycline (see *Treatment*). Also used are *E.coli* vaccines given to breeders to protect their chicks, e.g. Avicolivac (Rhône Mérieux)	Chlortetracycline hydrochloride 0.5 g per 5 litres in drinking water as Aureomycin soluble (Cyanamid) or neomycin as Neobiotic Soluble Powder (Upjohn) up to 0.5 g per 5 litres, or furazolidone up to 400 g per tonne of feed, or Furazolidone Soluble, e.g. Furazolidone Suspendable (Micro-Biologicals) 16 g per 5 litres of drinking water.
Fatty liver and kidney syndrome	Ensure protein and carbohydrate ratio is correct and also sufficient provision in feed of biotin and other vitamin B elements.	Administer water-soluble vitamin preparations suitably fortified with biotin, e.g. Duphasol 13/6–2 (Solvay Animal Health) 1 g per 4.5 litres of drinking water.
Fowl pox	Vaccination effectively prevents, e.g. Poxine and Poxinet (Solvay Animal Health) or Fowl Pox or Chick VI Pox (Vineland), or Diftosec CT Fowl Pox (Rhône Mérieux).	Administer non-specific broad spectrum antibiotics for secondary infections.
Fowl typhoid and pullorum disease (salmonella gallinarum and salmonella pullorum)	Blood testing and culling reactors on a regular basis combined with improved hygiene and husbandry. Vaccination can be used for typhoid.	Potentiated sulphonamides or furazolidone or broad spectrum antibiotics can be used, e.g. Duphatrim (Solvay Animal Health), Synutrim (Peter Hand Animal Health), Clamoxyl (SmithKline Beecham), Furazolidone BP (Peter Hand Animal Health).

Parasites and Disease	Prevention	Treatment
Gangrenous dermatitis	Usually associated with broilers after Gumboro disease challenge so vaccination against this assists. Do not use 'old' litter or over-stock and ensure efficient disinfection between crops.	Soluble penicillin in drinking water, e.g. Penbritin (SmithKline Beecham) 5.5 g per 200 litres of drinking water, or other soluble antibiotics with a broad spectrum of activity, such as amoxycillin (Amoxinsol 50 – Univet).
Gumboro disease (Infectious bursal disease)	Protect immature birds with living vaccines, e.g. D78 (Intervet Laboratories) Poulvac Bursine 2 (Solvay Animal Health), BUR 706 (Rhône Mérieux), Bursal Disease Vaccine (Vineland). To give parental immunity to chicks give inactivated vaccine to parents, e.g. Maternalin (Pitman-Moore) or NobiVac Gumboro (Intervet Laboratories).	Treat for secondary infections with antibiotics or antibiotic/vitamin mixtures.
Histomoniasis (Blackhead)	Emtryl Premix (Rhône Mérieux) 500–670 g per tonne of feed for turkeys or 340 g for chickens, or Sulfuride 14.6% (Solvay Animal Health), 345 g per tonne of feed for turkeys.	Dazole 40% Soluble (Peter Hand Animal Health) 67 g per 100 litres of drinking water. Emtryl Soluble (Rhône Mérieux) 30 g per 45 litres of drinking water or for short period use double dose.
Inclusion body hepatitis	Gumboro vaccination may assist in prevention as Gumboro disease appears to predispose to infection of IBH.	Non-specific measures including broad spectrum antibiotics and vitamins, especially fat-soluble vitamins.

Infectious avian encephalomyelitis (epidemic tremor)	A variety of live vaccines given in drinking water at an age of about 15 weeks, e.g. AE Vaccine Living (Solvay Animal Health), Delvax AE (Mycofarm), Avivac AE Vaccine (Live) (C.Vet), Tremblex (Vineland).	No specific treatment but encourage appetite and drinking by adding vitamin electrolyte mixtures to water.
Infectious bronchitis	Vaccination of immature birds by use of live vaccines, e.g. H-120 and H-52 as a coarse spray or in drinking water. Adults may be vaccinated by injection with inactivated vaccines. Examples are: Iblin inactivated or Iblin bivalent for laying birds (Pitman-Moore), or IB Vaccine Nobilis Ma5 (Intervet Laboratories), or Poulvac Infectious Bronchitis Vaccines H-120 or H-52 (Solvay Animal Health).	To curb secondary infections use chlortetracycline, furazolidone or potentiated sulphonamides. Examples of treatments for secondaries are: oxytetracycline: Terramycin Soluble Powder 5.5% (Pfizer); chlortetracycline: Aureofac 100 Feed Additive 20–50 mg/kg body weight (Cyanamid); Aureomycin Soluble Powder 20–50 mg/kg body weight (Cyanamid).
Infectious coryza (Roup)	Improve environment and hygiene and use vaccine, e.g. Haemovax (Rhône Mérieux)	Haemophilus sensitive broad spectrum antibiotics with supportive administration of soluble vitamins and electrolytes.
Infectious laryngo-tracheitis	Vaccination by individual ocular or nasal drop, e.g. Fowl Laryngo-tracheitis Vaccine (C.Vet) or Infectious Laryngo-tracheitis Modified Vaccine (Solvay Animal Health), or Laryngo-Vac (Intervet) or Fowl Laryngo-tracheitis Vaccine (Vineland).	Non-specific broad spectrum antibiotics and soluble vitamins and electrolytes.
Infectious stunting syndrome (ISS) (malabsorption syndrome)	No specific vaccines available but some benefit may result from the use of Reo- and Adeno-viruses combined with improved cleaning, hygiene and disinfection.	Administer generous quantities of soluble vitamins and electrolytes to assist absorption of essential nutrients.

Parasites and Disease	Prevention	Treatment
Insect infestation	A wide range of materials exist for dealing with flies, lice, beetles and crawling insects which are especially liable to infest poultry houses due to their warmth and their cavity construction. Nuvanol-N for beetles (Ciba-Geigy). Use 0.5–2 g Active. Kilsect (Turbair). Oil-based formulation with Turbair Flydowner Sprayer, 10 ml for 10 seconds per 10 square metres of surface for residual control. Flymort (Upjohn). Add 25 g to a litre of water and apply to affected area. Microgen (Micro-Biologicals). Fumigate at rate of 450 g per 500 m^3 of building. Long Life (Antec International) for terminal disinfection and disinfestation – dilute 1/320 to 1/60. Microcarb (Micro-Biologicals). 227 g per 27 litres of water. Apply at rate of 4.5 litres per 10 square metres.	
Marek's disease	Vital to keep chicks away from infection for first weeks of life. Vaccinate chicks at day-old and again at 10–21 days by intramuscular route, e.g. Marexine THV (Intervet) 0.1 ml per bird at day-old and up to 3 weeks. Marexine (MD) (Intervet), Marexine THV/CA (Intervet) or Cryomarex (various strains including Rispens) (Rhône Mérieux), or Poulvac Marek CVI (Solvay Animal Health).	No specific treatment but vitamin and mineral preparations added to drinking water will assist birds.

Mites	Disinfection of buildings and equipment with agents with established activity against mites. Removal of wood and other parts where mites are harboured.	Use only acaricides that are approved for direct contact with birds by spray, dust or dips.
Mycotoxicosis	As source is almost invariably contaminated feedstuffs, contents should be monitored and, where risk exists, add a chemical deterrent to feed, e.g. Bio-Add (BP Chemicals).	Little benefit from any medicine but soluble vitamins and minerals will assist recovery.
Necrotic enteritis	Improved hygiene including phenolic disinfectants and penicillin in feed at a level of 40 g per tonne.	Soluble penicillin in water at 0.5 g per 5 litres of water, e.g. Pensol (Micro-Biologicals) or broad spectrum antibiotic, e.g. oxytetracycline hydrochloride soluble up to 1 g per 4 litres, e.g. Terramycin (Pfizer).
Newcastle disease (fowl pest)	Variety of vaccines protect, e.g. Hitchner B_1 and La Sota by spray droplet or water medication, largely for young or growing stock, and inactivated oil-adjuvant vaccines by injection. Primarily for layers or breeders, e.g. Poulvac Hitchner B_1 or La Sota (Solvay AH), ND Vaccine Hitchner B_1 (Living) Nobilis, Newcavac Inactivated ND Vaccine (Oil Emulsion) (Intervet), plus numerous combinations.	No treatment effective against the virus of Newcastle Disease but treatment for secondaries as under *E.coli* Infections.
Perosis	Ensure a balanced ration, sufficient especially in manganese, other minerals and especially available phorphorus.	As prevention, administer extra soluble vitamins and minerals to assist utilisation of feed.

Parasites and Disease	Prevention	Treatment
Roundworms	Normal hygienic measures during and between batches of chicken on litter.	Piperazine hydrochloride. Mebendazole: Mebenvet 5% (Janssen).
Salmonellosis	General prevention relies on good hygiene and security of poultry sites, ensuring hygienic feedstuffs, and serology on flocks with elimination or treatment of those infected. After a site is infected it is esential to destroy all possible vectors of disease, i.e. rats, mice, litter beetles and insects.	Soluble furazolidone, e.g. Furazolidone Suspendable (Micro-Biologicals), 16 g per 5 litres of drinking water, or Penbritin (SmithKline Beecham) 5.5 g per 200 litres of drinking water, or neomycin, e.g. Neobiotic Soluble Powder up to 0.5 g per 5 litres of drinking water, or sulpha drugs.
Turkey rhino-tracheitis (TRT) of turkeys and swollen head syndrome of chicken	Vaccination, e.g. Aviffa-RTI (Rhône Mérieux) given from day-old onwards when flock is at risk.	Generally concentrates on secondary infection, e.g. furazolidone or potentiated sulphonamides, or broad spectrum antibiotics in feed or in drinking water.
Viral arthritis (tenosynovitis)	Live and inactivated vaccines available for good protection, e.g. Tenosynovitis Vaccine (Live) (Vineland), and Avian Reovirus Inactivated (Vineland) and Avian Reovirus Vaccine Inactivated (Rhône Mérieux).	Benefits are derived from the use of broad spectrum antibiotics, soluble vitamins and electrolytes.

12 Turkeys

Turkeys are no longer eaten only on special occasions, they are now a true round-the-year attraction to the consumer and very good value indeed. There are a variety of turkeys produced for the market. The 'mini' turkey, finished at about 12 weeks, weighs 4.5 kg (10 lb), or the medium weight ('midi') at 16 weeks weighs 6.5 kg (14.5 lb). Then, for the more traditional festive or, more likely, catering trade the heavier bird is required which will be kept from 18–30 weeks and will weigh from 8 kg (18 lb) to 15 kg (35 lb). In addition, the industry is marketing turkey portions, turkey rolls and turkey sausages, thus supplying the consumer with a very wide range of turkey products. The world production of turkeys is tending to rise quite consistently year by year. The smaller birds are kept rather like broilers in intensive housing, though usually not with such a fully controlled environment, whereas the latter may be reared in simple housing of the pole-barn type or even outdoors.

Unlike the domestic fowl, which almost invariably breeds naturally, modern strains of turkey are so cumbersome and broad breasted that natural mating is at least difficult and often impossible. Use is therefore made of artificial insemination, the males being 'milked' for semen to inseminate hens approximately once every two weeks, but often more frequently and even weekly. Most of the modern hybrids are based on two foundation breeds, the Broad Breasted Bronze and Beltsville White.

Rearing systems

Turkeys are brooded and weaned off heat at 8 to 10 weeks of age under intensive conditions, following which they may continue the growing period of 24 weeks or more in confinement, or they can be moved to good clean pastures to encourage their natural grazing habits. Short term rearing from day-old to 16 weeks is almost always undertaken in controlled environment houses. Breeding stock may be reared extensively or semi-intensively but though the capital cost of an enterprise of this sort is cheap, running costs are high and may be uneconomic.

Separate quarters are recommended for different age groups and the 'all-in, all-out' system whereby one age group is reared to killing age and the premises depopulated and cleansed is as strongly advocated for turkeys as for domestic fowl.

The principal brooding systems involve the use of either solid or wire floors. Floor brooding on a litter of wood shavings or peat moss requires a large floor area which must be free from draughts and the litter must be maintained in a dry condition to minimise the risk of parasitic infestations. Well-insulated and ventilated houses are required with supplementary heating to maintain a house temperature of 25°C (77°F) during the early stages of brooding.

Tier brooding cages are an alternative and once popular system, being especially made to give adequate head room and unobstructed access to food and water troughs, taking 50 to 70 poults to 3 weeks of age. They involve higher capital outlay than is needed for floor brooding, but allow greater concentration per unit of floor space. They are easily serviced and may be installed in many conversions. Less costly are frames of weldmesh fitted in a small unit or compartment of a house.

It is well to emphasise that early mortality in turkey poults due to lack of drinking or feeding is a constant problem that has to be prevented by good management and by providing the best equipment giving readily accessible food and water and plenty of bright lighting.

For the weaning stage, cooler battery cages may be used in a room temperature of 16°C (60°F) declining to 10°C (50°F). Alternatively an outdoor static hay-box fold with a stronger wire floor above a droppings pit will take 35–40 poults to 8 weeks old. Wire floor verandahs added to wire floor houses provide a further increase in floor space. The use of wire floors dispenses with litter problems, prevents access by the poults to their droppings and permits free circulation of air around the poults, but the complexity of the wire floor systems, together with high labour requirements, makes them increasingly unattractive.

Turkeys react to changes of environment and care is needed when these are made, especially when the change is extreme – such as a move from a wire or slatted floor to litter. In general, simple and standardised rearing on the floor is replacing other techniques and disease control is achieved by vaccines and drugs.

Growing and fattening range systems

Some flocks are still reared on range. Clean land which has not carried poultry or turkeys for two years is essential and is best divided into paddocks of 500 turkeys with ring fencing some 3 m high as protection from predators. Turkeys are extremely hardy when the feathers are 'bloomed' and on range they need only simple housing or shelters. Open or covered roosts and straw bale shelters will suffice. Some flocks may find security in a wire compound or farm building. Feeding and drinking equipment needs protection from weather, farm animals, wild birds and predators. Though more expensive to operate, the fold units of 20–40 turkeys provide better control with more even grazing and manuring of the land.

Intensive systems

These must provide adequate fresh air and floor space for growing stock. These requirements are met at low cost in the 3 m high wire netting or straw bale turkey compound, with open-air perches. Food and water troughs are placed outside and the compound area is covered with straw or wood shavings, but in a wet season this system is costly to operate. The turkey yard has the same basic arrangement but the perches are under a lean-to or farm building. The pole shed is a completely covered compound constructed of rough poles, railways sleepers or timber posts carrying a roof of wire netting overlaid with building felt or corrugated roofing material. Ends and back walls are fully or partially protected with straw bales, polythene or other sheeting but the remainder is covered with large mesh netting. Water troughs and supplementary food troughs are accessible from the front. No perches are provided and a 13 × 13 m section usually accommodates 300–500 turkeys.

Verandah and wire cage systems are both used for off-the-ground rearing in small units on limited land. Many of the modern units used for large scale rearing of part grown turkeys have efficient insulation, mechanical ventilation, supplementary heating and lighting to provide a controlled environment in all seasons and it is this arrangement, similar to that used with the chicken, which is gradually replacing all other arrangements.

Bearing in mind the rapid growth rate of turkeys the minimum space requirement shown in Table 12.1 should not be economised on any further and more space is always advantageous.

Poults grow rapidly and in consequence need increased headroom under the brooders, floor space, feeding space and ventilation in the house. It should be recognised that the poults at 4 weeks are 12 times their day-old weight and at 8 weeks are 3 times their 4-week weight.

After the young turkeys are hardened at 6–8 weeks and probably

Table 12.1 Space requirements (per bird)

Brooders	Hover units		0.008 m^2
	Suspended heaters		0.013 m^2
	Battery brooders		0.016 m^2
Intensive housing	Floor space	0– 4 weeks	0.08 m^2
		5– 8 weeks	0.12 m^2
		9–14 weeks	0.24 m^2
		15 weeks to killing	0.36–0.48 m^2
Pole sheds		15 weeks to killing (hens)	0.24 m^2
		(stags)	0.40 m^2
Compounds			0.48–0.72 m^2
Range	Clean pasture		450 turkeys per hectare

Feeding trough space	Minimum per 100 turkeys	
2 weeks	4 m	
3–4 weeks	6 m	With tubular feeders use
5–6 weeks	7 m	2–3 per 100 birds each
7–16 weeks	9 m	holding 13.62 kg of feed.
17 weeks to killing	12 m	

Drinkers	Minimum per 100 turkeys
0–4 weeks	3 × 2 founts
4–8 weeks	2 m
8–24 weeks	3 m

moved to new growing quarters the aim should be to maintain unchecked growth. At this time turkeys go through a stage known as 'shooting the red' when the head parts become a bright colour as the bird matures, and if the hens and stags have not already been separated, this is a good time to do it.

Turkeys to be reared in confinement should have their beaks trimmed to prevent vice, just prior to moving, but it is best not to make any changes in management until the birds are settled. They are most satisfactorily kept in small groups of up to 250 birds, with adequate floor and feeding space for their rapid growth in the next 3–4 months. Where perches are not used a few bales of straw will help prevent the young turkeys from herding and smothering when kept in large groups.

In close confinement a high standard of stockmanship is demanded since vices are more prevalent and disease risks are increased. Cleanliness at feeding and drinking sites and a liberal supply of clean litter is essential. Poor results are often due to overcrowding and inadequate ventilation which seem to be common errors in management in spite of all the recent evidence of their harmful effects.

Breeding turkeys

Breeders are maintained in two ways. One system is to maintain them on the floor in one of the several ways discussed earlier; it may be appreciated that controlled environment housing is gaining in acceptance because with its control of all conditions, including lighting, as with breeding chicken it has enabled much more efficiency and productivity to be achieved. Hens and stags may be kept together for natural mating or hens and stags may be kept separately and insemination may be achieved artificially. An alternative arrangement is to house the hens and stags all as individuals in cages kept in a house with a well controlled environment. Such a system gives maximum control of the birds and easier handling and is a system that continues to gain in popularity.

The following minimum areas should be provided but there is much to be said for increasing the space given to the birds if at all possible.

Hens and stags maintained on the floor in groups for natural mating	20 kg/m^2
Stags on the floor for artifical breeding	1 m^2 per bird
Hens in cages	30 kg/m^2
Stags in cages	1 m^2 per bird
Birds out of doors	20 m^2 per bird

Feeding

Management of turkeys is similar to that of the domestic fowl as regards housing environment, lighting, hygiene and disease control, but feeding is so important that the differences should be carefully noted (see chapters 3 and 4). The slowness in getting young poults to accept their food is well known and they must be coaxed to eat by making sure they can easily see the food and by having plenty of feeding points.

The same applies to water. Bright lighting is essential. All growing systems supply food *ad lib*, except where there is added grain which is given at special times in measured amounts. With outdoor methods no more food should be given than can be cleared up fairly soon after feeding as otherwise wild birds or vermin will consume it.

Turkey poults are supplied with food containing at least 24 per cent of crude protein for the first 8 weeks of life, reducing this level to 20 per cent at 12 weeks and 16 per cent at 16 weeks. A finishing ration for the heavier bird will have a protein percentage of about 13 per cent.

Substances similar to those in chicken rations are used. A poult 'starter' ration is fed up to 8 weeks and a 'growers' ration, which may be supplemented with grain if desired, is fed thereafter. The finishing ration is usually introduced about 3 weeks before the birds are to be marketed. Emphasis must be placed on the importance of insoluble grit for turkeys, as for chickens, especially when fed a grain diet.

The maintenance of health

The first stage of vulnerability for the turkey is at the very beginning of its life. They are much more inclined to perish than chicks or ducklings for no apparent reason, though it should be emphasised that they have very good sight. Thus, they need to be surrounded by water and food, plenty of light and adequate warmth – at least 35°C (95°F) at a house heat of about 25°C(77°F). It is desirable that this temperature difference of 10°C (22°F) between the brooder area and the house is maintained.

The intensification of turkey production has led to a considerable increase in the disease incidence. The chapters on the prevention and treatment of disease give adequate coverage of the diseases and should be studied carefully if health is to be taken seriously. What must be stressed here is that prevention is as always the best approach and can start very soon after hatching by administering vaccines – including those preventing Newcastle Disease, turkey rhino-tracheitis, pox, pasteurellosis and erysipelas. Whether some or all of these are necessary will depend on the local incidence. If rearing is to be done in an area where there are many turkeys, full vaccinal protection may be required, but in an isolated location little or none may be necessary. This illustrates only too well the importance of choosing the right area to keep turkeys though, sadly, poultry producers, for logistical reasons, can rarely be persuaded away from their suicidal tendency to bunch together – not unlike the birds they keep!

At present, the worst disease affecting turkeys is turkey rhino-tracheitis and is the latest viral infection to become a scourge for turkeys throughout most of the world. It has a really devastating effect on flocks of any age and after the initial viral damage, all the usual secondary viruses and bacteria can wreak great havoc. Losses of up to 30 per cent can easily occur accompanied by massive losses of productivity in those birds that stay alive. Treatment with medicines to reduce losses will usually help but sometimes hardly at all. Some help, or even a great deal, may result from improving the environment by providing more space, better air movement and ventilation and fresh litter. It should hardly be necessary to emphasise this again but I do not apologise for stating my belief that if we had no vaccines or medicines we could perhaps be even more successful in rearing turkeys, as with all livestock, by relying on good housing, hygiene, husbandry and management.

13 Ducks

With a tendency for the numbers still to be on the increase, the market for table ducks is steadily increasing. Egg layers are, however, declining as the strong flavour of the duck egg remains unattractive to the consumer. On the other hand, the table duck is being managed under increasingly sophisticated conditions which improve the value of the product for the housewife. Table ducks can make extremely fast growth, a live weight of 2–3 kg ($4\frac{1}{2}$–$7\frac{1}{2}$ lb) being attainable in 7–10 weeks at a food conversion efficiency of 3 : 1.

Breeds

The Aylesbury was by tradition the supreme table breed; it is a white feathered, light-boned duck with creamy white flesh, the adult drakes weighing about 4.5 kg (10 lb) and the females about 4.1 kg (9 lb). Its popularity has, however, been challenged by the Pekin, as a result of genetic improvement of this breed in the U.S.A. The Pekin is a smaller bird (about 0.45 kg (1 lb) lighter at the adult stage) but it is more fertile, a better layer and a more economical grower.

Two other table breeds are the Pennine, which has a good carcase quality and better egg production than the Aylesbury, and the White Table duck which is claimed to have a high percentage of meat. The two common egg-laying varieties, the Khaki Campbell and Indian Runner, are both prolific producers, an egg a day for a year being quite attainable. Modern hybrid strains of both egg and table birds are being developed from existing breeds. They have the same advantage as the chicken hybrids, in that they combine the best features of two or more breeds, yet still produce a high degree of vigour and uniformity.

Breeding

The duck appears to respond to the lighting stimulus in the same ways as the chicken – that is, breeding activity is induced by an incremental increase in the period of light in each 24-hour period.

Breeding pens should contain one male for every eight females; they should be made up in groups at least a month before eggs are required for hatching. Great care should be taken when designing accommodation to ensure that nest boxes can be kept clean, especially in view of the traditional hazard of duck breeding – infection with the *Salmonella* group of bacteria. The nesting material must be changed frequently and the eggs collected soon after laying, then washed in a combined detergent and antiseptic solution at a temperature slightly above that of the eggs. Thereafter the eggs should be dried, fumigated and stored at 14–15°C (58–60°F) and a relative humidity of 80 per cent.

Usually eggs are not used for hatching until the ducks have been laying for about 6–8 weeks. Incubation requires a temperature of 37.5°C (99.5°F) and a humidity of 75 per cent 31°C or 87°F wet bulb.

Feeding

Table 13.1 gives examples of three suitable rations for ducks through their life cycle. It is considered a preferable techique to feed all three rations in pelleted form.

In the U.S.A. growing stock are often fed rations containing higher protein levels (18–20 per cent). It is important to maintain the correct ratio between the energy and crude protein levels and the figure should be 140:1 for the starter feed and 160:1 for the growers.

A substitute feeding regime for the young stock is a starter diet for the first two weeks preferably in the form of crumbs, alternatively as a wet mash. During this time a trough space of 50 mm per duckling should be allowed. From two weeks onwards the 'grower' food is best given as a pellet and the trough space increased up to 100 mm per duck. Birds to be kept for breeding are given pelleted feed and often grain as a night feed with a total intake of 0.3–0.4 kg (6–8 oz) per day, which is similar to the food consumed by the adult. In the management of ducks it is most important that the food is given in troughs which are capacious enough to enable the birds to make use of the shovelling action of their beaks. Ducks of all ages drink large quantities of water and whilst this must be provided, vessels must be designed so they can be kept clean, to avoid bacterial infections.

Table 13.1 Examples of duck rations (kg per tonne mix)

	Starter ration 0–2 weeks	Grower ration 0–6 weeks	Finisher ration 6 weeks to finishing	Breeder ration	Holding ration*
Wheat meal (coarse)	500	423	600	433	800
Maize meal (fresh)	267	329	300	300	—
Barley meal (fine)	—	100	—	—	165
Fish meal	113	50	31	67	—
Soya-bean meal	100	54	—	67	—
Meat and bone meal	8	25	50	33	22
Grass meal	—	—	—	33	
Limestone (flour or granules	10	10	10	60	—
Dicalcium phosphate	—	7	—	—	—
Fat	—	—	6	—	—
Supplement (vitamin/mineral)	13	13	13	13	13
Calculated analysis					
Crude protein %	20.2	15.5	13.3	17.0	11.3
ME kcal/kg	2885	2877	2965	2694	2825
Ca %	1.24	1.12	0.99	3.1	0.21

* Suitable for breeding ducks between 8–20 weeks of age and for moulting between laying periods.

Brooding

Young ducklings can be brooded by floor and battery systems similar to those described for chickens but at slightly lower starting temperatures of 30°C (86°F). Three tier battery units measuring 1200 mm × 600 mm × 250 mm and holding 25 ducklings to 3 weeks of age are frequently used, but many growers prefer a single tier arrangement. A unit measuring 1050 mm × 2100 mm will take about 100 ducklings for the first week of life. Ducklings brooded on the floor can be kept in a variety of buildings, the traditional system being a long narrow house with pens on each side of a centre passage. With this arrangement the ducklings can later be released for the fattening period into an outside yard running on each side of the building.

Litter materials include straw, sawdust, shavings, peat moss and sand, but the large quantities of water that ducks drink and therefore void present a severe problem with almost any form of bedding. That is why a wire base, as in tier brooders and follow-on cages and simple wire floored systems, tends to be used for the brooding and subsequent periods. A suitable wire floor for the first 4 weeks consists of a 12.5 mm mesh of 8 gauge.

Rearing and fattening

This is the period from about 4 weeks onwards when ducks would traditionally be put outdoors. They are placed in groups of about 200 at a density of 5000 to a hectare and they should be on a light sandy soil such as is found in the local Breckland of Norfolk in the U.K. where a high proportion of the industry is based. The pens must be moved frequently and require a light fencing only about 1 metre high; shelter from the worst effects of the weather can be provided by wind breaks and covers of straw bales and corrugated iron. There are, however, serious doubts on the economics of outdoor systems, especially during the winter and an increasing amount of indoor fattening is carried out both in winter and summer. This is often done on straw bedding and at other times on welded wire or slats – 25 m mesh of 8 gauge being suitable and allowing up to 0.1m^2 of floor space per bird by the end of the fattening period. In indoor systems the ducks are frequently provided with access to outside yards of asphalt, concrete, or merely well sanded earth. These appear to have considerable benefits in that they give the birds more room at low cost, and they can certainly be advocated provided that the base is well drained and (if a permanent floor is not used) measures are taken to rest and disinfect the soil. Wire-floored verandahs can also be incorporated with all indoor housing systems, being a cheap and safe way of providing fresh air and exercise.

Mention should be made of one or two other less common arrangements. For the rearing of breeding birds colony housing may still be used, the ducks being reared in a small pen at one end of a house and moved into larger pens as they grow. Grass pens in conjunction with slatted floor houses are also used for small groups, either for rearing or breeding. Occasionally hay-box brooders in which the ducklings' warmth is maintained and retained in a nest of hay, rather than by the use of artificial heat, are used for the period from 2–5 weeks.

The space allowances are shown in table 13.2. They can be taken only as a general guide to the needs of the birds. It is never wrong to give more

Table 13.2 Space allowance per duck(ling) – indoors

Age (weeks)	Space allowance (m^2)
1–3	0.045
3–4	0.068
4–5	0.09
5–6	0.113
6–8	0.18

and this is especially true in small groups. Outside a space allowance of up to 1 m^2 is a reasonable figure.

Adults

A variety of arrangements of equal success can be used for housing adult breeding birds. Contrary to common belief there seems no virtue in providing swimming water, but cool *ad lib* drinking water is necessary.

The simplest method of all is to put the laying flocks on free range at a rate of about 250 ducks to the hectare. If the pasture is reasonably good this enables some economies in feed and the ducks can benefit from access to ground, such as marshes, which could not support other livestock. While adults can be kept outside without any form of protection this is not advisable economically, particularly in winter, and there is also an increased health risk. Housing can, however, be simple timber or hardboard construction with a floor of well-drained concrete or wire mesh. An allowance of 0.2–0.3 m^2 per bird is sufficient.

Various other semi-intensive or intensive arrangements may be used. A simple system seen commonly in Europe consists of rows of narrow pens each containing about 8–10 breeding or laying birds, the ground being sloped to one end and usually having a sandy base. A small shelter is provided at one end.

Nowadays, however, the trend is towards the maintenance of adult breeding stock or laying stock in controlled environment intensive housing similar to that used for chickens. A suitable system is to have the house divided into slatted areas which contain the drinkers and a bedded area, usually of straw and shavings. Housing may even be heated in an attempt to reduce the amount of straw used by increasing the drying effect of the air. Ventilation rates in intensive housing must also be maintained at a high rate and automatic fan systems are favoured when stocking rates are high. For intensive housing small flocks of about 200–250 birds are recommended. Nest boxes should be provided at the rate of one nest to every three ducks. They are usually placed on the floor around the walls or against the partitions between pens.

In order to obtain a constant supply of ducklings throughout the year two arrangements may be employed. Either the ducks are induced to produce hatching eggs for as long as possible – up to about 25–30 weeks – or they are allowed to lay for about three months, then force moulted and rested for a similar period, after which there will normally be another laying period. Moulting can be induced in several ways, by moving the birds, by feeding a grain-only diet, or by gradually reducing the lighting periods.

In the successful management of ducks good housing, coupled with high standards of nutrition, are the cornerstones of a profitable programme but hygiene is especially important as a supporting procedure. Traditional systems, in the main, are not satisfactory under the more intensive regimes now favoured and improved standards of health control and environmental regulations are tending to increase indoor management and reduce the following of the outdoor arrangements.

Maintaining good health

Unlike chickens or turkeys, few vaccinations are normally required for ducks. Vaccination against duck virus hepatitis (DVH) (see also chapter 11) may, however, be advisable. DVH vaccine should be used only on farms which have experienced a previous incidence of the disease and the same is true when vaccines are used for duck virus enteritis (DVE) and pasteurellosis.

We rely very heavily on good 'quality control' to keep disease away. Ducks quickly make an area into a mess with a strong likelihood of a build-up of infection occurring. The desirability of giving generous space allowances, good underfoot conditions, a healthy environment and constant attention to hygiene, cannot be exaggerated. If conditions are bad there may be an upsurge in fungal infections, which are difficult to treat, and in breeders egg-borne disease can create enormous losses. There is also the possibility of ducks being infected with parasites, especially when they are on pasture. The effect of parasitic infection, as with bacterial infections, may be sub-clinical so that no obvious signs of disease are seen but nevertheless there will be heavy losses due to poor productivity. If the duck owner has the slightest indication that his birds are so affected, a sample of mortality and/or live 'culls' should be submitted for examination to a veterinary laboratory or it is even better if the veterinarian can visit the premises and examine all relevant aspects.

14 Geese

The production of geese has changed little over many years and there has been little attempt at genetic improvement. The common types in this country are English grey, English white and English grey back. They are based essentially on the original breeds of Embden and Toulouse with an admixture of others. Whilst there are considerable variations, ganders reach weights of 8–9 kg and a goose weighs about 7 kg. Egg production varies between 30 and 80 eggs in a season.

Breeding

Selection for breeding should be made on the usual basis, that is, egg production, fertility, hatchability, viability and conformation. It is important to make up the breeding pen (or 'set') of geese with care as they will not mate and breed satisfactorily unless they are in harmony with each other and have had some weeks living together. It is advised that the sets are made up about two months before the season's breeding in the spring. They can then stay together for years; there are plenty of records of geese living together and breeding for two decades at least. The mating ratio is best at 4 or 5 geese to 1 gander though a wider ratio is not infrequently used. Each set should be allocated to a separate house when it is made up and confined within a run separated from other sets. Thereafter, once established, a number of sets can be run together in a paddock.

Incubation

Goose eggs require between 29 and 32 days' incubation. They may be hatched naturally, either under a broody goose or a hen. It is a reasonably simple procedure – the broody 'mother' is put in a separate coop on an earth floor and bedding of straw or hay and fed and watered regularly.

The eggs should be turned once a day and the nest have a little warm water run under it to ensure the eggs do not get too dry.

If the eggs are incubated artificially, the machine should be run at the same temperature and humidity as chicken eggs and turned not less than once a day. A goose egg will take up the space of about four chicken eggs. Procedures for 'candling' eggs and removing infertiles should be followed in the same way as with incubation generally.

Brooding

If geese are brooded naturally the goose or hen 'mother' will remain with them as long as required.

If reared artificially, similar facilities can be provided as for turkeys, chicken or ducks. A temperature of 38°C (100°F) is the normal starting figure, whether the goslings are kept on wire floors or on solid floors with litter. As the droppings from goslings tend to be rather wet it is vital for their good health to ensure that the litter is tended and cleaned or topped-up to keep the condition healthy and dry. Artificial heat will not be needed for more than three weeks but the goslings should be let out onto short grass as soon after hatching as possible. Whilst it is advisable to keep the birds in a fairly restricted area for the first week or two, they may soon be given real 'free range'. The only housing required is a basic shelter with a sound dry floor. It is a benefit if the birds have access to a pond but it is not an essential requirement.

Nutrition

The basic nutritional requirement for the profitable farming of geese is to provide quality grass, short, succulent and nutritious. It is not a good

Table 14.1 Approximate live-weight for age

Age	Weight (kg)
Day-old	0.10
4 weeks	1.30
8 weeks	2.80
12 weeks	3.70
16 weeks	4.50
20 weeks	4.90
24 weeks	5.50
28 weeks	6.50

procedure to put the geese on long fibrous grass as this has little value.

At the beginning of their life the goslings may be started on duck crumbs or mini-pellets. This is better than the more traditional mash feeding though it can be used if preferred. This food may be provided *ad lib* for the first few weeks and should have the addition of some small size insoluble grit. The birds must not be overstocked. A reasonable maximum stocking rate for permanent pasture is around 100 geese to a hectare. However, a more fruitful and intensive way of rearing the geese is to stock them much more heavily – up to 600 birds per hectare – but to keep moving them around at 4-weekly intervals to new ground. Geese for the table will be ready at 10–12 weeks, especially if good quality supplementary feed is provided. After the goslings have reached 8 weeks of age, approximately, the mixture may be usefully supplemented with some whole grains. The feeding of the breeders depends on quite different factors. During the winter months the birds will need to be fed a simple maintenance diet, usually a mixture of wheat, barley, maize and dried grass. It is desirable to give the breeders a diet supplemented with adequate vitamins and minerals. It is not easy to be dogmatic about the quantities of food used but one can keep a careful watch on the condition of the geese. Ensure that all birds have access to the litter.

Table 14.2 is a sample of Ministry of Agriculture UK-approved and recommended diets for 'home mixers'.

Table 14.2 Goose diets (kg/tonne)

	Starter	Grower	Breeder
Wheat (coarse ground)	300	350	250
Barley (fine ground)	200	300	200
Maize (fresh ground)	300	200	300
Dried grass	50	—	—
Meat and bone meal	20	30	60
Fish meal	50	30	60
Soya-bean meal	60	70	80
Limestone (flour or granules)	10	10	40
Dicalcium phosphate	5	5	5
Salt	5	5	5
Vitamin/mineral supplement	12	12	12
Calculated analysis			
Crude Protein	16	15.6	18
Energy kcal/kg	2765	2788	2733

The future

There is a limited demand for geese and production has actually been in decline until quite recently when there has been some sign of a revival. The fall in demand has been due to the poor quality of the products with poor quantity of meat and excessive fat. Nevertheless the meat has a most attractive 'natural' flavour and if more attention is paid to the correct feeding and breeding the profitability could return and the rearing of geese could form a useful adjunct to a mixed farming operation.

As with the production of quail, improved productivity can be achieved by applying our increasing knowledge on poultry production in general. For example, geese are traditionally seasonal layers but investigations have shown that out of season production can be stimulated by the use of a 16–17 hour day. A period of short days (6–8 hours) needs to be given to the breeding geese before the 16-hour day for a good effect.

Geese are relatively free of disease but they should be checked for parasites, especially the gizzard worm, and duly treated with Levamisole. Management techniques can reduce the chance of such diseases by reducing the stocking density or moving the geese more frequently. Improving selection by better breeding techniques and hygiene and hatchery practice can also reap considerable dividends. None of these proposals will compromise the good flavour and natural methods that commend the geese for the table.

15 Quail

Quail production throughout the world is on the increase both for meat and egg consumption. Whilst they are still considered to be largely luxury foods, improvements in the genetic quality of the birds has led to better productivity and a more satisfactory economic outlook for their more widespread demand. The future of quail production indeed looks encouraging world-wide.

The most commonly used breed, especially in the United Kingdom, is the Japanese quail (*Coturnix coturnix japonica*). The average female weighs about 250 grams, the male a little less. Egg weight is some 14 grams. The quail lay some 200 eggs in a year and the young birds are ready for the table at about 40 days at a live weight of 250 grams.

Housing and environment

Quail can be grown either in cages or on litter. Whilst the traditional practice is to brood in one area and transfer to a growing house at 2–3 weeks, it is really much better not to have the unnecessary stress of a move in the quail's short life. Whichever system is used, the arrangement and techniques are similar to those for chicken and brooders; cages and housing and equipment are adapted. The correct ambient temperature is vital for good quail production and the birds should be brooded at 35–36°C (95–97°F) gradually reducing over 3 weeks to 20°C (68°F). A good start is vital and it is best to commence with feed and water in plastic apple trays, weaning them on to automatic cup or nipple drinkers shortly afterwards. The birds should be floor brooded at 150 birds per square metre, reduced at 2 weeks to 100 birds in the same space. If the birds are reared in cages they can be housed at a rate of double that provided for the chicken. Light intensity is most important and it should be very bright to start with (60–70 lux), reduced to about 10 lux by 3 weeks where lighting is fully controlled. The birds can be given 23 hours light a day, as it is best to give them an hour's darkness from day-old so they will be accustomed to a power or lighting failure and will not panic if plunged into darkness.

With birds reared on the floor the best litter is soft wood shavings or chopped straw. Special care must be taken to see that the litter has been stored in a dry state since dampness can cause aspergillosis, and this respiratory disease can easily cause havoc amongst quail since they are more susceptible than most birds. If it is preferred to rear in cages, ordinary chicken tier brooders may be used at a density of about twice that used for chicks; special tier brooders and follow-on brooders are also made for quail.

Feeding

Quail being reared for meat are usually given *ad lib* turkey rations. Most growers use a turkey super-starter ration from day-old to 14 days and thereafter there is a change to turkey starter to the finish. The best form of feed is a small crumb, especially small in the first stages of the birds' life and smaller than is usual for turkey poults or chicks. By 40 days the birds will be near mature weight and this is the time to kill. They will have a food conversion efficiency of approximately 2.2 from hatching to finishing. The feeding of breeders requires a different régime. The birds require a diet of 20 per cent protein and 11.5 MJ/ME/kg. In practice those types of ration given to a light hybrid chicken breeder will be satisfactory, though some farmers provide a proprietary game breeder's ration. The birds will require approximately 25 g per head per day. As there is a limited demand for quail eggs for human consumption, some birds will be kept for this purpose and can be given the same feed as light hybrid chicken layers.

Breeding

Selection of breeders is usually made on the basis of inspection, weight and conditions. The main requirement is to have birds in perfect health and showing good uniformity. The birds should be obtained from known recommended lines with good egg laying ability, viability and hatchability. It is most important to avoid inbreeding during the operation of a breeding programme. Using criss-cross breeding, taking males and females from alternating families, inbreeding is avoided and a measure of hybrid vigour is maintained yet good characteristics are retained. It is wise to keep a minimum of six lines and preferably more to assist with this programme.

When making up the breeding pens the best procedure is to have 2 or 3 males with 6 to 9 females. Such a pen of no more than 12 birds is very satisfactory but it is currently more common to place the birds in larger units. The females should keep up production for the best part of the year. Fertility will fade, however, quite markedly after 5–6 months, but the position can be improved by replacing the males at about 5 months in the breeding pens. In the breeding pen, as quail are hot weather birds, the temperature should be maintained at least at 20°C (68°F). Lighting schedules can be similar to those advised elsewhere for breeders.

Incubation

Quail eggs are extremely delicate and must be handled with great care. Many breeders prefer to dip them in a sanitising solution before storage which should be between 16–20°C (59–68°F) and a relative humidity of no less than 75 per cent. They should not be stored for more than 7 days.

The eggs can be incubated in the same machines as for chicken eggs but the equipment will take double the number. Eggs take just under 16 days from setting to pipping and a further 10 hours from pipping to hatching. The régime advised is as follows in the machines. At the start the temperature should be 37.5°C (99.5°F) and the relative humidity 87 per cent (wet bulb temperature 30.6°C (87°F). At about 15 days the temperature should be reduced to 37°C (98.6°F) and the relative humidity increased to 90 per cent (32.3°C wet bulb temperature). The temperature should be changed to 37.2°C (99°F) at take-off and the relative humidity to 80 per cent (27.6°C wet bulb temperature). Eggs must be turned up to 7 times a day if it is not an automatic turning machine.

Traditional housing arrangements

For the small and traditional grower the housing designs shown in Figs 15.1 and 15.2 are strongly recommended by an undoubted expert on quail, Mr G.E.S. Robbins (see details in *Further Reading*). The quail breeder house (Fig. 15.1a) is easily constructed from timber and oil-tempered hardboard for outside use. The food and water are replenished without upsetting the birds. In the breeder house small nest boxes are placed so that access for collection of the eggs is through the small hinged flap shown in the diagram. Also shown are simple quail pens (Fig. 15.1b) in which the quail can be grown on the floor and a recommended design is also shown of a range of quail aviaries (Fig. 15.2).

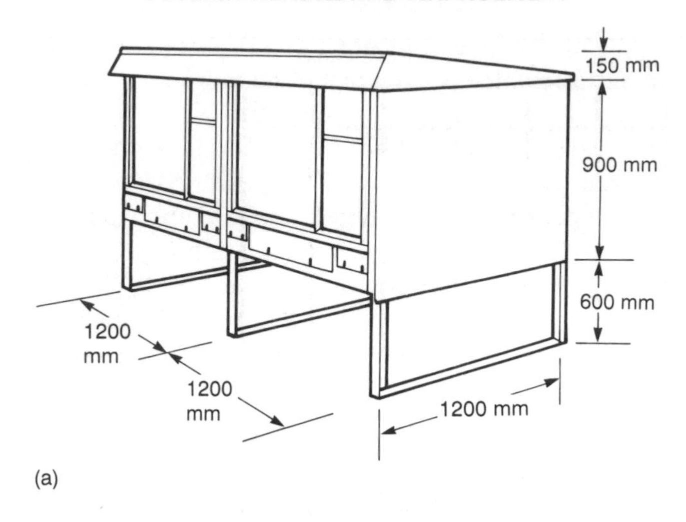

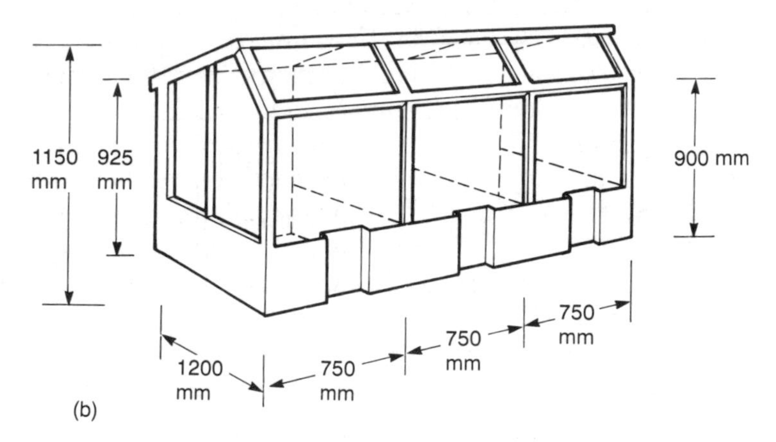

Fig. 15.1 Designs for (a) quail breeder house and (b) quail pens.

Health and productivity

Quail are affected by a range of diseases which is very similar to other poultry but under most conditions with good hygiene and management losses are small as enterprises are generally modest in size. Some of the commonest conditions are listed below (for details on treatment consult the earlier chapters on diseases and Table 11.3):

- *Respiratory conditions* due to *Aspergillus fumigatus*, mycoplasma and secondary infections – very similar to chronic respiratory diseases of chickens, including pasteurella.

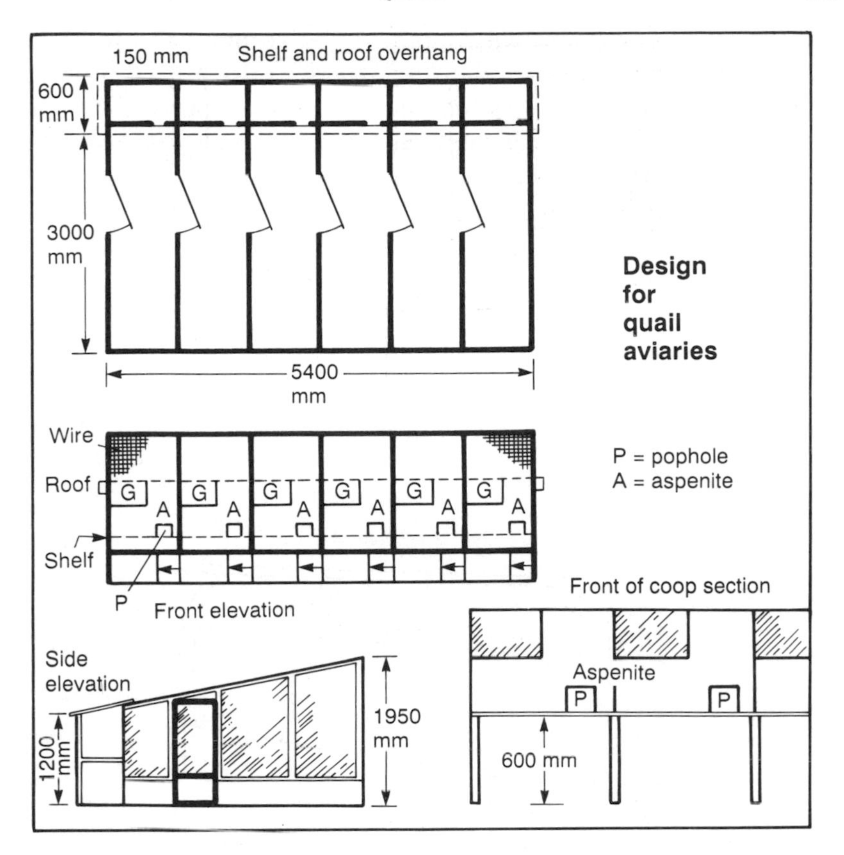

Fig. 15.2 Design of a range of six Quail aviaries constructed by Keith Howman.

- *Intestinal disorders* due to bacterial build-up, the commonest being clostridia, *E. coli* and staphylococcus.
- *Skin disorders* such as mites are common under bad hygienic conditions.
- *Parasitic conditions* may also occur ranging from coccidiosis to blackhead, intestinal and bronchial parasites.

The future for quail

The market on the continent of Europe and especially in France shows that there is a good potential for specialist markets and if productivity can be improved these interesting prospects are opened up for the future. There is no reason why very great advances should not be achieved by applying better breeding techniques, enhanced methods in the hatchery and improved environmental conditions for keeping the breeders and growing the young birds to maturity. There is also a need to improve hygienic measures in line with those used in other sections of the poultry industry as most of the diseases that occur are avoidable by better management and methods.

16 The management of poultry in hot climates

Large poultry units are being increasingly developed in areas of high temperature that are not traditional to advanced methods of husbandry and special techniques should be considered for the satisfactory management of birds under these conditions.

Housing and environment

Naturally ventilated (or convection housing)

There is no firm agreement amongst experts as to whether it is preferable to have open types of freely ventilated natural convection housing, or controlled environment housing, in hot climates. My own preference is to use naturally ventilated housing with relatively light stocking densities wherever there are doubts about the reliability of the power supply, or the management skills are relatively unsophisticated. Buildings such as those shown in the diagrams (Figs. 16.1 and 16.2) will give ideal conditions in most warm climates and there is a great facility for variation in the amount of ventilation and air movement by the use of controllable curtains at the front or sides of the building. This is generally essential as diurnal variation in temperature and air movement can be very great in tropical climates.

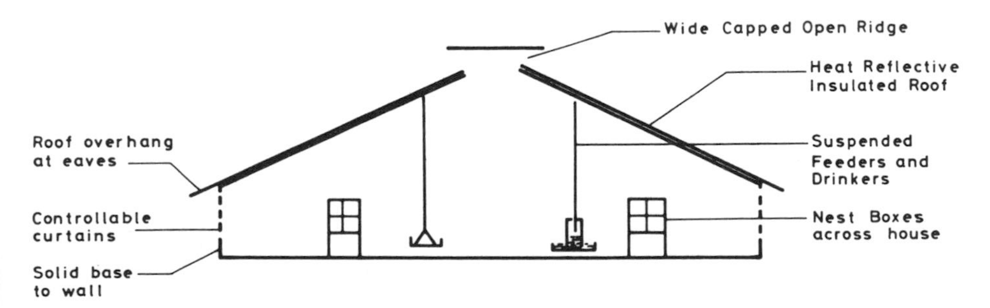

Fig. 16.1 Span building for birds in hot climates.

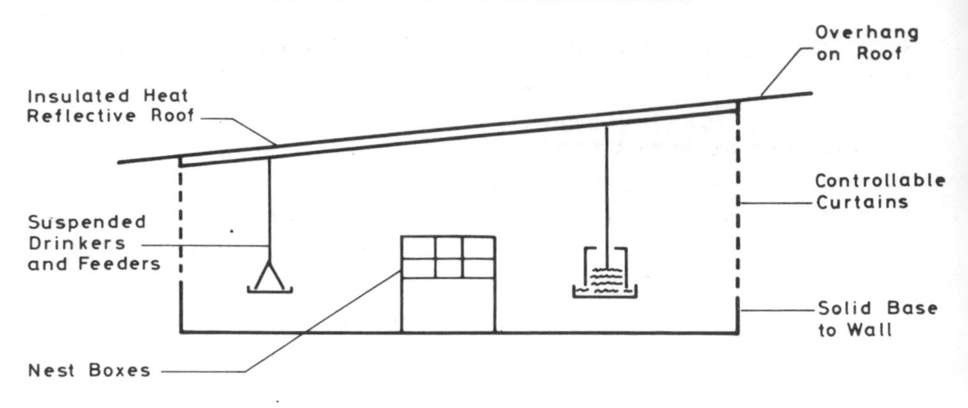

Fig. 16.2 Monopitch building for birds in hot climates.

In nearly all circumstances it is desirable that the roof of the building is insulated to a high standard – equivalent to about 150 mm of glass fibre or 100 mm of expanded plastic. The outer claddings should be in light heat reflective materials.

The width of naturally ventilated houses should not exceed 13 m (40 ft), and the ridge height should be up to 4 m (13 ft) with a generous opening at the ridge up to 0.6 m (2 ft) suitable capped to prevent rain entry. A sharp angled pitch to the roof will help ventilation to be effective. There should also be an overhang on the roof beyond the eaves of up to 1 m (3 ft) to give protection from the heat of the midday sun. The curtains at the front or side should extend over 70–80 per cent of the area and can, if desired, be controlled automatically by thermostats and motors with emergency 'fail-safe release' for high temperatures. Stocking rates for these type of buildings would be no more than about 7 broilers per m^2 (15 kg body weight per m^2) or 4–5 layers or 3–4 breeders per m^2 depending on the size of the birds.

High air speeds are of great benefit to the birds and the attendant as they increase the heat loss by convection. Nevertheless it must be stressed that these speeds should be higher than are normal in poultry houses and there is some danger that if the temperature goes above 41°C, which is the body temperature of the bird, high air speeds will *increase* the stress on the birds. Because of the importance of being able to increase the air movement whenever it is needed in naturally ventilated housing it must be possible to open up as much as possible of the walls of the house without restriction and I would not hesitate to advise incorporating the design ability to open up 100 per cent of the sides in very warm areas. Careful thought must also be given to the siting of naturally ventilated buildings so the best use can be made of winds to flow through the house when required.

Controlled environment housing
This system is especially advocated for heavy weight broiler breeders or heavy broilers which cannot withstand very high temperatures because of their bulk and therefore poorer heat dissipating abilities. It is also widely favoured for large cage laying houses where natural ventilation presents too many circulation difficulties. Stocking densities can be between 50–100 per cent more than naturally ventilated housing and the principal features of controlled environment housing are as follows.

A high standard of thermal insulation is the first essential. The whole building should be insulated to a standard of 100 mm polyurethane or the equivalent in other materials as chosen. Mechanical ventilation should be installed to a maximum of approximately double that in temperate climates, and given earlier in chapter 7 on ventilation. Whilst all the systems illustrated in chapter 7 can also be used in hot climates, there is a general preference for positive pressure arrangements, as they enable a better control of air movement to be achieved and this is of especial importance in hot climates. In order to help the circulation of air it is also quite common practice to install ceiling circulating fans which may improve the environment in comparatively 'dead' areas with little air movement. Reliable electricity supplies and stand-by generators must be available.

It is noteworthy that building structures with a large thermal mass reduce the diurnal variation in temperature and in this way lower the extremes of environmental temperatures experienced by the birds. To put it more simply, such buildings are slower to warm up in the day or to cool down in the night. In many parts of the world suitable local materials are available that can be used ideally for this purpose, for example, light-weight clay bricks or forms of aerated concrete. In buildings so constructed, the potentially large diurnal swings of temperature may also be further reduced by increasing the ventilation rate during the night and thereby cooling the whole structure of the house and then reducing the ventilation from its maximum in the heat of the day so that the building materials have a better opportunity of remaining cool. It would be especially appropriate to use one of the water cooling methods, described later, during the day to further assist the levelling out of temperature fluctuations. The temperature stresses that tend to kill poultry are caused largely by suddenly rising temperatures as the birds have had no ability to acclimatise themselves and the method just described will be a significant help in preventing the most serious temperature stresses that kill the bird.

The main systems for keeping the birds cool use methods of evaporative cooling, of which there are four principal forms. The simplest is the *pad or filter system* in which the incoming air is drawn over a type of filter, often of a cellulose material, which is kept soaked in water. Provision has to be made to protect the pad from excessive sand or dust and recirculation and filtration of the water is also usually necessary.

A second system, also quite simple, is to use a series of *low pressure fogger nozzles* in the house, set up close to the ordinary fans or incoming air, to assist in the distribution of the mist. This system tends to become quickly clogged and easily leads to uneven conditions with the litter near the fogger becoming dangerously damp.

There are two superior arrangements. The first is by the use of a *spinning disc* to generate a spray of small droplets which are passed into a stream of air. The most popular of the disc coolers are horizontal free standing units. However, a more recent development is the *ultra-high pressure mister* (Fig. 16.3). In this case a pump is used to generate a high pressure in reinforced plastic distribution tubes fitted with special nozzles. The small droplet particles pass out into the house at about 100 metres per second. The water supply must be carefully filtered and the pipes distributed in the house so that the incoming air will pick up the mist and help in the distribution.

The various cooling systems given here are capable of lowering the ambient temperature in the house by a maximum of about 10°C (16°F) when relative humidities are down to around 30 per cent, but if the

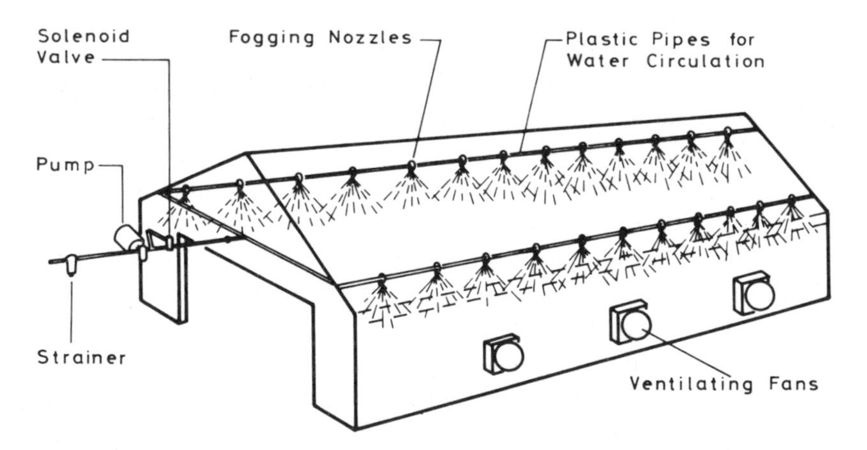

Fig. 16.3 Illustration of a high pressure misting (fogging) evaporation cooling system in a poultry house, by courtesy of NRG Company, PO Box 306, Ardmore PA19003, U.S.A.

relative humidity is up to 50 per cent the effect will be nearer 6°C (10°F), with a rise in relative humidity to 80 per cent.

A method much used where water is freely available is to trickle water over the roof of the house from perforated plastic pipes. The evaporation of the water will certainly have some beneficial effect and it also has the advantage that this system does not raise the humidity of the air inside the house. It tends to be a method most favoured in those temperate climates where water is freely available and only occasionally are high temperatures met with. It is also cheap and easy to install a perforated plastic pipe along the ridge of the house and the occasional extravagance in water usage is a modest cost compared with the potential loss of birds suffering from heat stroke in large numbers.

It is worthy of emphasis that high temperatures *per se* are by no means the only problem. High relative humidities, over about 70 per cent, are extremely serious and birds can perish in substantial numbers even at quite modestly high temperatures if they are given an abruptly severe heat stress. Without acclimatisation, heavy birds in particular are very susceptible to such conditions at high stocking densities.

Before poultry units are set up in potentially dangerous climates, the meteorological data should be carefully examined to establish whether the whole operation can be economically justified.

Feeding and watering in hot climates

Since birds will eat to their energy needs and their energy requirement is obviously less in hot climates, it is essential to compensate for the inevitable reduction in total food consumed. For example, with pullets being reared for laying or breeding, the protein content of the ration, which is usually at a level of 20 per cent to 4 weeks, then 15 per cent for 4 weeks and finally 12 per cent until the pullet is laying, should be amended to a level of 20 per cent for the first 6 weeks and then reducing only to 16 per cent and leaving it at this higher level until the pullets are housed for lay. It is also necessary to increase the vitamin and mineral content of the feed to the rearing bird and, as food consumption may be up to 20 per cent less at high temperatures, the vitamin and mineral levels should be increased to compensate by a similar margin of 20 per cent to give a safe, but neither harmful nor wasteful, quantity. In addition there should be an especially generous supply of drinkers – about 20 per cent more space should be provided than in temperate climates. Because of the danger of early dehydration the first water provided for the young chick should have electrolytes and vitamins added to increase its attraction and value to the chick.

It is of benefit if the birds can actually dip their beaks in the water so that drinkers with troughs are better than nipple systems. One of the ways in which birds keep cool is by losing heat from the comb and wattles and this ability will be greatly improved if they are kept moist. Only if there are fairly deep troughs can the birds thoroughly moisten these parts.

At high temperatures birds may consume twice, or, exceptionally, three times, as much water as at the normal range of temperatures. Unless this water can be properly provided, there will not only be insufficient for the physiological functions of the bird but there may also be a serious reduction in feed consumption. For each kilogram of feed the bird may drink up to 3 kilograms of water. In addition, the cooler the water that can be provided, the better, and water storage facilities should therefore be protected from excessive radiation effects from the sun.

The food for all other ages and classes of birds should receive the same consideration – assuming about 20 per cent less consumption, and increasing the protein, mineral and vitamin percentages by an approximately pro rata amount.

In order to encourage food consumption in hot climates when controlled environment housing is used it is worthwhile giving the hours of darkness during the heat of the day and lighting the birds during the cooler night time. In fact not only will this encourage food consumption, it will also help to reduce the heat stress on the birds by encouraging them to rest during the day.

Some important additional points must be made. Deterioration in some of the essential micro-ingredients at high temperatures may readily take place and since transportation of feed ingredients over long distances and very uncertain storage conditions are almost inevitable, it is advised that if there is uncertainty about the quality, analyses should be carried out to check the position. Furthermore, under such uncertain conditions the growth of moulds producing pathogenic mycotoxins is more likely than in cool temperatures and it becomes essential to ensure the best feasible storage conditions or consider the possibility of incorporating an anti-mould agent in the food. In any event an examination for mycotoxins in the food may be necessary.

Poultry diseases in hot climates

The health position tends to be worse in hot climates than temperate areas. There are several reasons for this. First and foremost, there is not the tradition for hygiene in many such areas and yet paradoxically pathogenic micro-organisms can often survive for a longer time in a warm environment – this is particularly true of some of the most devastating bacterial infections caused by salmonella, pasteurella and clostridia. Several strains of these organisms also cause diseases that are communicable to man, such as food poisoning and gangrenous infections. Also, in hot climates there are greater infestations of the vectors that cause disease; biting insects, flies and mites may abound and these can be carriers and reservoirs of many diseases, especially those of viral origin. Usually there will also be considerable deficiencies in the immune protection chicks have derived from their parents so that they may receive serious challenges from disease organisms as soon as they go into the house at day-old.

Provided these dangers are understood, proper measures can be taken to prevent their worst effects. In addition to all the environmental and nutritional points already made, the following extra measures should be considered.

(1) Maintain the cleanliness of the site at the highest standards ensuring no accumulations of litter are left near the birds as in these the vectors of disease may live and thrive.

(2) Take the greatest care to keep visitors away unless they are fully protected by antisepsis and clean clothing.

(3) Apply a fully integrated programme of preventive vaccination which is tailored to local knowledge of the disease incidence.

(4) Protect the birds when necessasry at times of stress with antibiotic medication for short controlled periods under veterinary direction.

(5) Exclude vermin and wild birds from the housing by suitable wire mesh over all ventilation and other open areas.

(6) Dispose of dead carcases well away from the site or, best of all, by incineration.

It is as well to bear in mind that disease and environmental challenges will be much reduced if the birds are given more space than in cool and temperate areas.

17 Poultry welfare and alternative systems

It has already been stressed in earlier sections of this book that there is considerable world-wide concern about the possible inhumanity involved in some of the modern poultry production systems. The objections have arisen for two principal reasons – the restriction on the movement of birds that systems such as the battery cage impose, and the barren nature of the birds' surroundings, with their whole lives spent squatting on a wire floor and being restricted by wire mesh or metal. Though considerable investigation is currently under way to try and establish the birds' mental and physiological reactions to environments such as the battery cage, no results so far appear to have been clear-cut enough to give us a definite answer. The most satisfactory approach at present is perhaps epitomised in the Codes of Practice revised by the U.K. Farm Animal Welfare Council. It is said that all farm animals should be provided with the following:

1. Comfort and shelter.
2. Freedom from thirst, hunger or malnutrition.
3. Freedom of movement.
4. Prevention or rapid diagnosis and treatment of vice, injury and disease.
5. The company of other animals, particularly of like kind.
6. The opportunity to display most normal patterns of behaviour.
7. Adequate lighting.
8. Flooring which neither harms the animal nor causes undue strain.
9. The avoidance of unnecessary mutilation.
10. Emergency arrangements to cover fires and breakdowns.

A great debate will doubtless continue for many years on the definition and interpretation of these clauses but under modern intensive systems the principal difficulty is with item 6 – the opportunity to display most normal patterns of behaviour. For chickens this could well be said, as many insist it does, that the bird must be able to stand up, sit down, turn round, flap its wings, have dust baths and lay its egg in a bedded nest.

Alternative systems for the management of laying poultry

The cage system of housing laying poultry is without any doubt the most economical method of producing eggs. Furthermore, cage systems have regularly become more intensive by a steady increase in the number of tiers of cages in battery blocks. Arrangements with as many as eight tiers have been built. However, although the productivity and health of birds are better than other systems, there are serious welfare disadvantages in the cage system. Cages not only fail to provide the all important welfare requirements of freedom of movement, freedom from fear, comfort and shelter, suitable flooring and freedom to display most normal patterns of behaviour but also lead to certain metabolic problems such as bone fragility.

Since cages for poultry were first criticised by animal welfare groups some 30 years ago there have been continuous endeavours to develop alternative systems which can provide a satisfactory level of welfare and also competitive productivity. These systems, however, have incurred problems of poor productivity, increased disease incidence, intensive management, aggression and cannibalism. All such systems are more costly than cages as a method of producing eggs. An outline of these methods is given below.

1. The get-away cage (Fig. 17.a)

The first welfare-orientated cage was produced about 17 years ago and

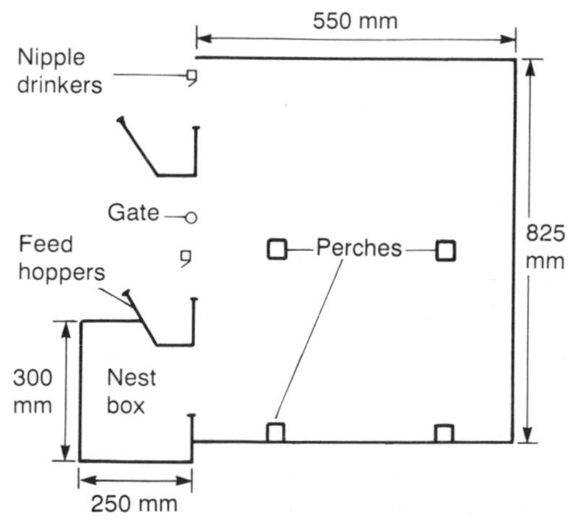

Fig. 17.1 Get-away cage (Elson 1976).

had within its confines perches, nest boxes and 'dust baths'. There are a number of alternative arrangements of these. Such designs have found little following due to their expense and some management problems. However, they are being further developed, e.g. with time-monitored opening and closing of nests and sand baths.

2. The Elson tiered terrace (Fig. 17.2)

A more recent system developed by Dr Arnold Elson of the United Kingdom Agriculture and Development Advisory Service is a tiered terrace which aims to combine good control of the birds and 'high' technology of cages with the allowance of more freedom for the birds. Tiered sloping wire mesh floored terraces are interconnected by a 'stairwell' giving controlled access in the afternoon to a littered ground floor area. Feed, water, nesting and perches are provided on each terrace, so that birds return to them at night via one way gates. So far testing has shown that the production can be good but cannibalism may be a problem.

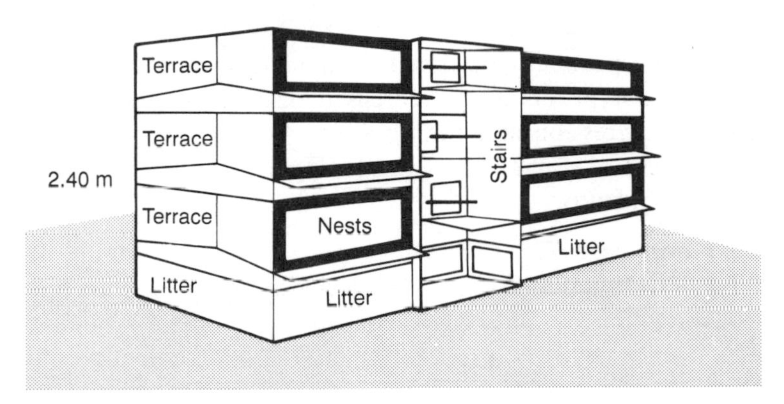

Fig. 17.2 Elson tiered terrace (ETT system).

3. The aviary system (Figs. 17.3–17.5)

There are a number of aviary systems; these were the first of the multi-tiered alternatives to the cage which gave freedom of movement for birds within an intensive housing arrangement. The multi-tier arrangement approximates more closely to a bird's natural habitat. Temperature, ventilation and lighting are controlled and the vertical height of the house is made use of by allowing the birds to live at a number of different levels on slats, perches or wire floors. In addition there is always a littered area and nest boxes. The problem faced with this sytem are similar to those in the perchery arrangement described below.

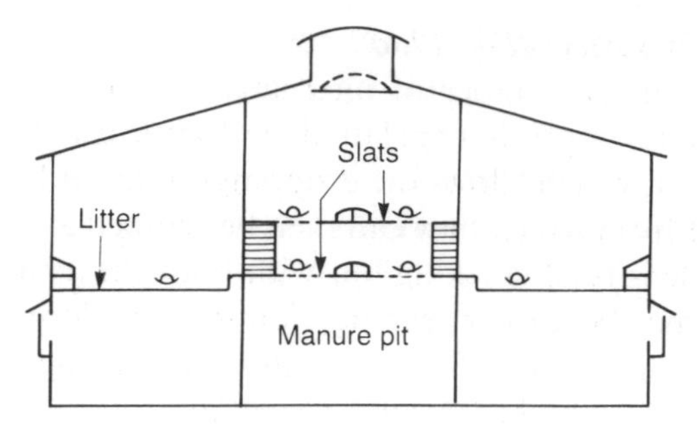

Fig. 17.3 Aviary at Gleadthorpe, England.

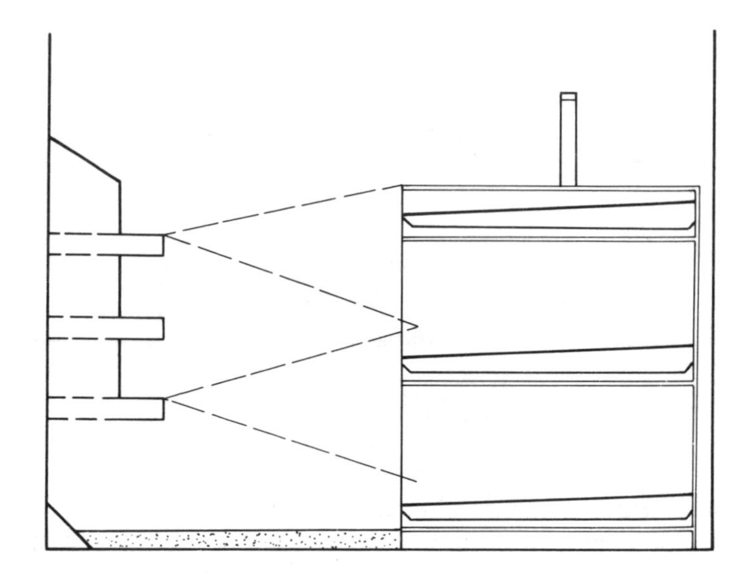

Fig. 17.4 Swiss aviary – 'Natura 280'.

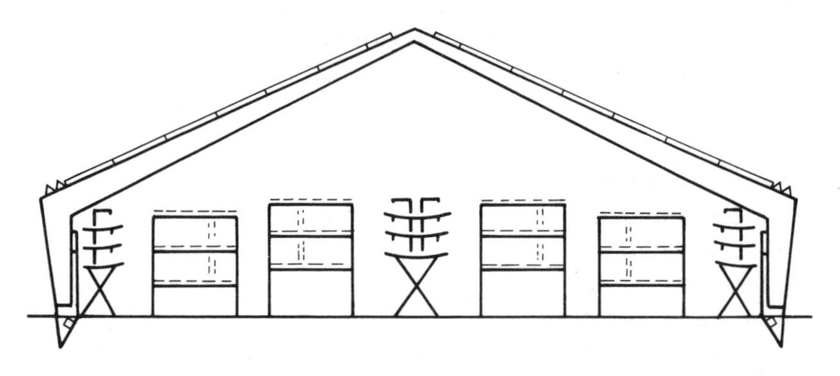

Fig. 17.5 TWF aviary (Netherlands).

4. The perchery system (Fig. 17.6)

Though there are a number of alternative arrangements, the system consists of a series of perches or slatted platforms, usually on an A-frame over a wire floor, which allows the droppings to fall either into a deep pit or on to a belt from which they can then be scraped away. Such systems allow a high density of stocking but with both the aviary and perchery systems there may be feather pecking, cannibalism, floor and broken eggs and high disease incidence. Figure 17.6 shows an arrangement of this type of system pioneered by Dr Michie in Scotland.

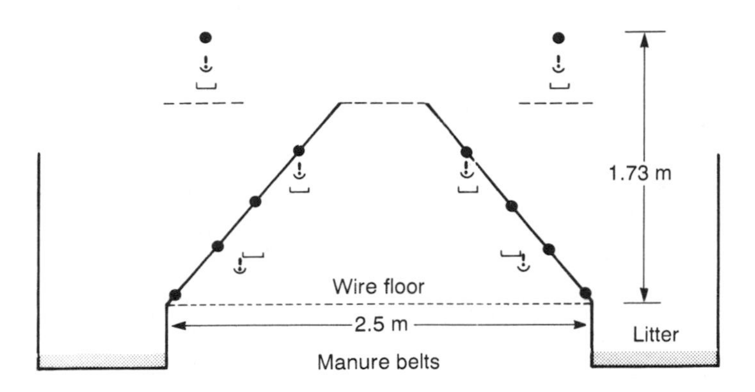

Fig. 17.6 Michie perchery (Scotland).

5. Sloping wire floors (Pennsylvania systems)

A variety of sloping wire floor systems has been developed from the original Pennsylvania system where the birds were confined to a sloping wire floor on which they lived their entire life and also laid their eggs. Modern developments may include nest boxes, perches and sanded scratching areas. The customary behavioural problems occur – cannibalism, feather pecking and egg-eating. One such system is that produced by Dr Hans Kier (Figure 17.7) and is being used in Denmark.

6. Tiered wire floors

Further developments of the original sloping wire floors use several levels of floors, incorporating automatic belt collection of manure, and perches and littered floors for scratching and dust bathing. These are often high density systems with birds housed at 20 birds per square metre and could show considerable promise if the behaviouristic problems in high stocking rates can be overcome.

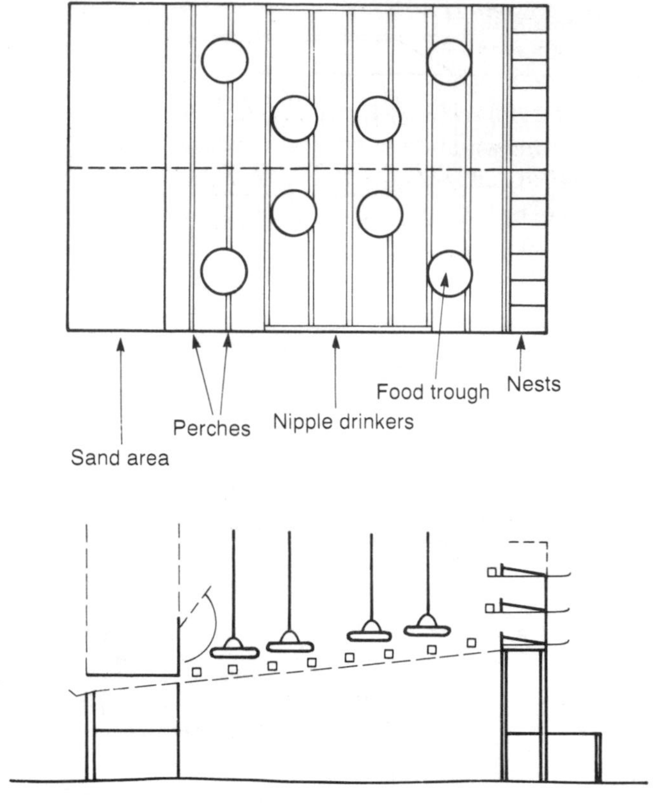

Fig. 17.7 Plan and cross section of the Hans Kier system (Denmark).

7. Deep litter

The deep litter system was first developed in its modern form some 50 years ago and with improvements is generally used for breeders. The house has built-up litter of wood shavings or chopped straw, a controlled environment, automatic feeding and drinking, nest boxes which may have automatic egg collection, and in many cases an area of perches over a droppings pit. The system is satisfactory provided stocking density is not excessive. It is an expensive system to erect and to manage and for these reasons is not used for commercial egg layers.

8. Covered straw yard (Fig. 17.8)

Birds are kept in a simple covered yard which is deeply strawed. Ventilation is by natural air flow and lighting likewise, supplemented only by artificial lighting when necessary. The straw litter is kept at a depth of about 500 mm and is constantly topped up to keep it at least to this depth.

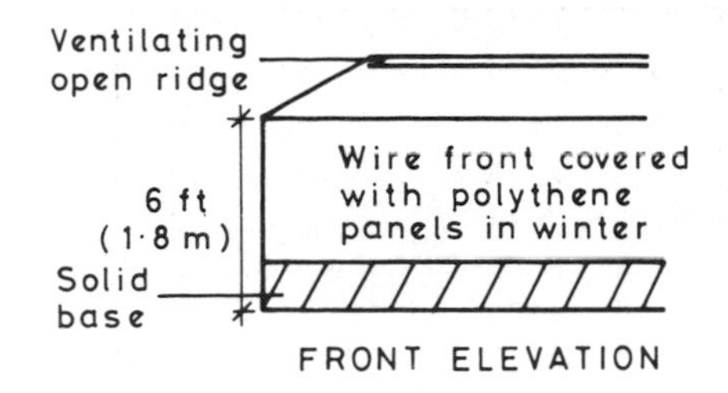

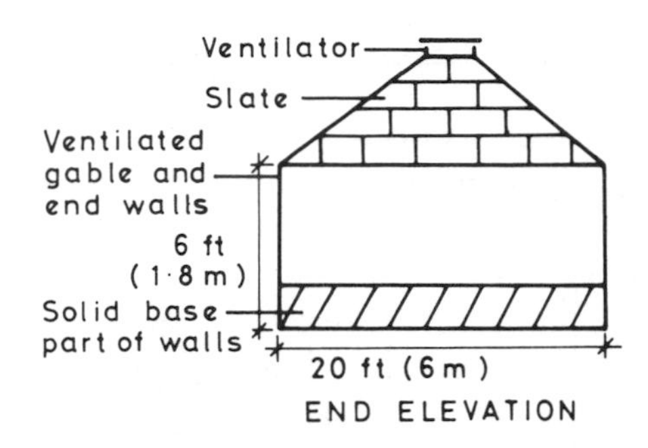

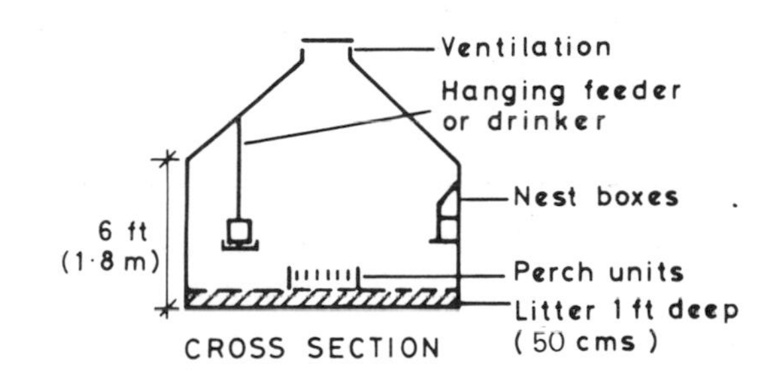

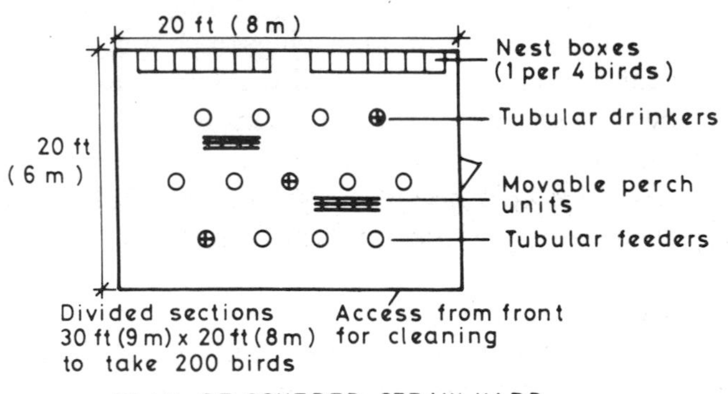

Fig. 17.8 Plans of covered straw-yard.

Food and water supply is automatic, egg production is good but feed consumption may increase due to absence of environmental control. Birds may roost on perches in a separate insulated area. The birds should not be stocked above four birds per square metre and should be kept in groups of not more than 200–300. Vices may develop if the birds are overstocked. The system is easy to run, especially if units are of modest size and the system has had a following for years.

9. Free range (Fig. 17.9)

This is a system where birds are given free access to pasture and for egg-laying poultry they must not be stocked above 1000 birds per hectare. The house itself is virtually the same as a deep litter or perchery house. The system is only economic because the return on 'free range' eggs is much better than eggs from cages or 'barn' eggs, which are those from the alternative systems. The system can run into problems from cannibalism, floor eggs and dirty eggs. Food consumption and labour demands are high. Disease can be a problem if the land is over used. Bad weather and predators may also cause difficulties. The highest standards of stockmanship are essential.

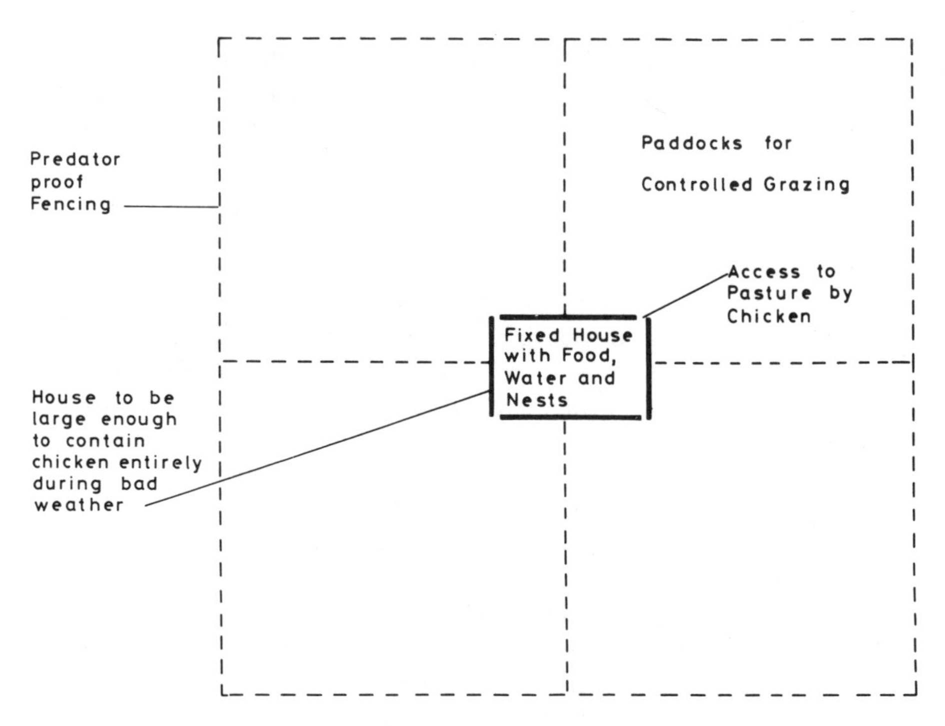

Fig. 17.9 A plan of modified free range chicken unit.

Table 17.1 Estimates of egg production costs in different poultry systems

System	Space	Cost
Laying cage	450 cm^2/bird	100
Laying cage	560 cm^2/bird	105
Laying cage	750 cm^2/bird	115
Laying cage	450 cm^2/bird + perch	100
Laying cage	450 cm^2/bird + perch and nest	102
Get-away cage and 2-tier aviary	10–12 birds/m^2	110 115
Aviary and perchery and multi-tier housing	20 birds/m^2	108
Deep litter	7–10 birds/m^2	118
Strawyard	3 birds/m^2	130
Free range	1000 birds/hectare	150

Indices are based on the percentage over laying cages stocked at 450 cm^2/bird.

10. Semi-intensive system

This is the traditional domestic poultry keepers' method, with a small fixed timber house leading to one or more open runs. The arrangement is only satisfactory for small groups of birds as it has serious environmental, labour, productivity, health and hygienic disadvantages.

Free range table chicken

An increasing number of table chicken are now produced under free range conditions. There are a number of features which are fundamentally different from the intensive broiler bird. Instead of a fast growing broiler strain, the free range bird is red feathered, slower growing and maturing so that it takes nearly twice as long to reach a similar weight as a broiler. After the first two or three weeks of life brooded on straw or wood shavings the birds are given free access to pasture. The diet is generally a largely cereal one with the sufficient addition of vegetable protein, vitamins and minerals but no animal protein is added. Nor are growth promoters given, and the protein content and energy levels are several percentage points below broiler diets.

It is claimed that with the extra time taken to mature the flavour is improved and indeed a bird is offered with a more traditional chicken flavour. It also has much less fat than the broiler bird.

Conclusions

It is not easy to foresee clearly where the welfare controversy is going to lead. Research and investigation may eventually provide a scientific answer but in many ways the moral issue has tended to make the running and will probably overrule the scientific evidence. Our legislators, moved by public demand, are likely to consider the keeping of layers in multi-bird cages unacceptable, in which case it is good to know that there are a number of alternative arrangements that can be used and which will provide eggs of good quality at very little extra cost than those laid by the battery bird.

If, however, the alternative systems to battery cages are to be successful in all respects, including welfare, the poultryman must be aware that good results will not come easily and management will require greater skills. Some of the most popular trends affecting the evolution of the modern poultry industry will be abruptly halted or reversed and these must be considered in a practical manner.

Over several generations commercial egg layers have been expertly bred for their efficiency under cage conditions and there will have to be changes in genetic selection techniques if the ideal strains for future needs are to be produced.

Diseases are likely to be on the increase unless great care is taken. The birds can readily get infested with parasites and other agents of disease in most alternative systems unless the litter is kept in a dry and friable state and only modest stocking rates are allowed. There will also be a revived risk of vices occurring, such as feather pecking and cannibalism. There is unlikely to be any desire to trim the beaks of birds and the prevention of vices will depend entirely on skilled management.

Cages with sloping wire bases almost certainly ensure that eggs are clean with no need for effort on the part of the attendant, but it can be very different with floor systems. If all the birds use the nest boxes provided and these are kept clean there is no problem but, quite unpredictably, there may be a tendency for some or even many birds to lay their eggs on the floor. This has many dangers. The eggs will probably be dirty and at the worst are even a danger to health. Some eggs may get broken accidentally and this may set in train 'egg-eating' which can cause a serious loss of eggs. A great deal of investigation is currently being undertaken into the behaviour of laying birds in an endeavour to ensure that nests are correctly designed and sited to encourage birds to use them exclusively, but there are so many differences in the behaviour of various strains of birds that it is too much to expect a total success.

Repeatedly, it must be said that the final results depend on management. Just as the cage system has been the main factor which has encouraged the growth of the mammoth egg laying units, so may the development of alternative systems see a change to smaller groups of birds in smaller buildings and on smaller poultry units. Many may applaud such a development, which will not only be an aid to good management but may also make better use of the resources of the land. It is much more likely in the case of the smaller unit to utilise locally produced feedstuffs and the litter at the end of the crop will be of great value to the land. In large intensive units, on the other hand, the poultry feed may have to be hauled a great distance and the manure at the end of a crop is a liability. If we can also see the removal of some of the hazards inherent in systems that are totally dependent on mechanical environmental control and feeding, there can be improved management since time is spent on inspection of the birds rather than on the repair of the machinery.

Poultry husbandry is never a dull subject and nothing stays the same for long. At the moment we are at one of the most crucial and interesting stages in its development where welfare interests are having a profound influence on the direction poultry systems are moving.

Further reading

1. General management

Anon. (1989) *Poultry Serviceman's Manual.* New Jersey, U.S.A.: Merck, Sharpe & Dohme International.

Bartlett, T. (1990) *Ducks and Geese: A Guide to Management.* Swindon: Crowood Press.

Card, L. and Nesheim, M. (1982) *Poultry Production.* Philadelphia: Lea & Febiger.

Carter, F. (1986) *Intensive Poultry Management for Egg Production.* London: H.M.S.O.

Cole, D.J.A. and Brander, G.C. (1986) *Bioindustrial Ecosystems.* Amsterdam: Elsevier.

Ducks and Geese (1983) Bulletin No. 70, Ministry of Agriculture, Fisheries & Food (M.A.F.F.) London: H.M.S.O.

Gleadthorpe Experimental Husbandry Farm Reports (1974 onwards). London: M.A.F.F.

Robbins, G.E.S. (1989) *Quail, Their Breeding and Management.* Reading: World Pheasant Association.

Robbins, G.E.S. (1990) *Introduction to Quail in Captivity.* Reading: World Pheasant Association.

The Rearing of Pullets (1983) Bulletin No. 54, M.A.F.F. London: H.M.S.O.

Thompson, M.A. (1978) *The Organic Poultryman.* Bridport, Dorset: Matthew A. Thompson.

Universities Federation of Animal Welfare (1988) *Management and Welfare of Farm Animals.* London: Bailliere Tindall.

2. Breeding, hatching and welfare

Baxter, S.H., Baxter, M.R. and MacCormack, J.A.O. (1983). *Farm Animal Housing & Welfare.* The Hague: Nijhoff.

Jones, D.R. and Hodgetts, B. (1987) *Incubation and Hatchery Practice.* Bulletin No. 148, M.A.F.F. London: H.M.S.O.

Lake, P.E. and Stewart, J.M. (1978) *Artificial Insemination in Poultry.* Bulletin No. 213, M.A.F.F. London: H.M.S.O.

Romanoff, A.L. (1960) *The Avian Embryo.* New York: Macmillan.

Romanoff, A.L. and Romanoff, A.J. (1949) *The Avian Egg.* New York: John Wiley.

Sainsbury, D. (1988) *Farm Animal Welfare.* Oxford: Blackwell Scientific Publications.

3. Nutrition

Ewing, W.R. (1983) *Poultry Nutrition.* Pasadena, U.S.A.: The Roy Ewing Co.

Feltwell, R. and Fox, S. (1988) *Practical Poultry Feeding.* London: Faber & Faber.

McDonald, P., Edwards, R.A. and Greenhalgh, J.F.D. (1988) *Animal Nutrition.* (4th Ed.) Harlow: Longman.

Nutrient Requirements of Farm Livestock – Poultry (1975) London: Agricultural Research Council.

4. Environment and housing

Animals, Pathogens and the Environment – A Review. (1991) Paris: International Office of Epizootics.

Appleby, M.C. and Hughes, B.O. (1991) Welfare of Laying Hens. *World Poultry Science Journal*, 47(2), 109–128.

Charles, D.R. and Spencer, P.G. (1976) *The Climatic Environment of Poultry Houses.* Bulletin No. 212, M.A.F.F. London: H.M.S.O.

Esmay, M.L. (1979) *Principles of Animal Environment.* Westport, Connecticut, U.S.A.: AVI Publishing Co.

Hafez, E.S.E. (1968) *Adaptation of Domestic Animals.* Philadelphia, U.S.A.: Lea & Febiger.

Livestock Environment (1974) Proceedings of the International Livestock Environment Symposium. St Joseph, Michigan: American Engineers. Special Publication SP-0174.

Sainsbury, D. and Sainsbury, P. (1988) *Livestock Health and Housing.* London: Bailliere Tindall.

Woolcock, J.B. (1991) *Microbiology of Animals and Animal Products*, World Animal Science, A6. Amsterdam: Elsevier.

5. Poultry health

Curtis, P. (1990) *Poultry and Game Bird Diseases*. Liverpool: Liverpool University Press.

Dawson, P.S. ed. (1978) *Salmonellas and Poultry*. M.A.F.F. and British Veterinary Association.

Gordon, R.F. and Freeman, B.M. (1971) *Poultry Diseases and World Economy*. B.E.M.B. Symposium No. 7. Edinburgh: Longman.

Hofstad, M.A. *et al.* (1986) *Diseases of Poultry* (7th Ed.) Ames, Iowa, U.S.A.: Iowa State University Press.

Jordan, F.T.W. ed (1990) *Poultry Diseases* (3rd Ed.) London: Bailliere Tindall.

Merck Veterinary Manual (7th Ed.) (1991). Rahway, N.J., U.S.A.: Merck & Co. Inc.

Index